CAMBRIDGE TRACTS IN MATHEMATICS

General Editors

B. BOLLOBAS, W. FULTON, A. KATOK, F. KIRWAN,
P. SARNAK

144   Torsors and Rational Points

Alexei Skorobogatov

Imperial College of Science, Technology & Medicine

Torsors and Rational Points

CAMBRIDGE
UNIVERSITY PRESS

PUBLISHED BY THE PRESS SYNDICATE OF THE UNIVERSITY OF CAMBRIDGE
The Pitt Building, Trumpington Street, Cambridge, United Kingdom

CAMBRIDGE UNIVERSITY PRESS
The Edinburgh Building, Cambridge CB2 2RU, UK
40 West 20th Street, New York, NY 10011–4211, USA
10 Stamford Road, Oakleigh, Melbourne 3166, Australia
Ruiz de Alarcón 13, 28014 Madrid, Spain
Dock House, The Waterfront, Cape Town 8001, South Africa

http://www.cambridge.org

First published 2001

Printed in the United Kingdom at the University Press, Cambridge

Typeface Computer Modern 10/13pt   *System* LaTeX2e   [UPH]

A catalogue record for this book is available from the British Library

Library of Congress Cataloguing in Publication data
Skorobogatov, Alexei, 1961–
    Torsors and rational points / Alexei Skorobogatov
        p.   cm. – (Cambridge tracts in mathematics; 144)
    Includes bibliographical references and index.
    ISBN 0 521 80237 7
    1. Torsion theory (Algebra) I. Title.

    QA251.3.S62   2001
    512.4′–dc21   2001025430

ISBN 0 521 80237 7 hardback

Вижу, вижу, как в идеи
Вещи все превращены.
Те - туманней, те - яснее,
Как феномены и сны.

Возникает мир чудесный
В человеческом мозгу.
Он течет водою пресной
Разгонять твою тоску.

...

На кустах сидят сомненья
В виде галок и ворон,
В деревах - столпотворенье
Чисел, символов, имен.

Перед бабочкой пучина
Неразгаданных страстей...
Геометрия - причина
Прорастания стеблей.

Н. М. Олейников. "Пучина страстей" (философская поэма), Финал.

Contents

1

Introduction

A fundamental fact about Diophantine equations is that there can be no algorithm determining whether a given equation is soluble in integers $\mathbf{Z}$ or not. This is the famous negative solution of Hilbert's tenth problem by M. Davies, H. Putnam, J. Robinson, Ju. Matijasevič and G. Čudnovskiĭ. More precisely, there exists a polynomial $f(t; x_1, \dots, x_n)$ with integer coefficients such that there is no algorithm that would tell us whether for an integer t the equation $f(t; x_1, \dots, x_n) = 0$ is soluble in integers or not. The polynomial $f(t; x_1, \dots, x_n)$ can be made explicit, for instance, we can have $n = 13$ (see, for example, [Manin, L], VI).

In this book, however, we are mostly interested in the solubility of Diophantine equations in the field of rational numbers $\mathbf{Q}$ and more general number fields. In this case the analogue of Hilbert's tenth problem is still open. For homogeneous equations the existence of solutions in $\mathbf{Z}$ and in $\mathbf{Q}$ is, of course, equivalent provided one does not count the all-zero solution.

For certain classes of equations an algorithm deciding the solubility over $\mathbf{Q}$ can be found. Such is the case when a class of projective varieties defined over $\mathbf{Q}$ satisfies *the Hasse principle*. This principle consists in requiring that the obvious necessary conditions for the solubility of a system of homogeneous polynomial equations with integer coefficients $F_i(x_1, \dots, x_n) = 0, i = 1, \dots, m$, that is, the solubility of congruences modulo all the powers of prime numbers, and the solubility in the field of real numbers $\mathbf{R}$, be also sufficient. (The solubility of congruences modulo all powers of a prime number p is equivalent to the solubility in p-adic integers $\mathbf{Z}_p$.) If the Hasse principle holds, the problem of the existence of a rational point reduces to a purely local problem at finitely many critical places of $\mathbf{Q}$. The first non-trivial example when the Hasse principle holds is when we have one quadratic equation; this is the famous Minkowski–Hasse

theorem. However, very soon one encounters rather simple equations which
are *counter-examples to the Hasse principle*: these are equations soluble in
all completions of $\mathbf{Q}$, that is, $\mathbf{R}$ and the fields of p-adic numbers $\mathbf{Q}_p$ for
all primes p, but not in $\mathbf{Q}$. Such are the homogeneous cubic equations
$3x^3 + 4y^3 + 5z^3 = 0$ (Selmer) and $5x^3 + 9y^3 + 10z^3 + 12t^3 = 0$ (Cassels and
Guy), where one of course excludes the all-zero solution. The first of these
equations defines a smooth projective curve of genus 1, and the second one
defines a smooth cubic surface. Despite the failure of the Hasse principle
there are reasons to believe that for cubic equations in any number of variables there does exist an algorithm telling us whether a given equation is
soluble or not.

In the middle of the twentieth century, the Minkowski–Hasse theorem
motivated Mordell, Selmer, Châtelet and others in their search for other
cases of the Hasse principle and similar local-to-global principles, and in
the analysis of the cases when it fails. In the course of this were discovered
concepts of particular significance such as the Selmer group of an elliptic
curve, the Tate–Shafarevich group, and the Cassels–Tate form on it, and
finally, its vast generalization, the Manin obstruction to the Hasse principle.

It was Manin who found a first general obstruction to the Hasse principle, and a good substitute to the Hasse principle when it does not hold
is the statement that *the Manin obstruction to the Hasse principle is the
only obstruction*. This means that as long as a collection consisting of a real
solution and p-adic solutions for all primes p satisfies certain conditions,
there also exists a solution in $\mathbf{Q}$. These conditions, provided by the Brauer-
Grothendieck group of the variety, are based on the global reciprocity law.
To verify each condition requires finitely many elementary operations, and,
in the ideal cases, there are only finitely many such conditions. The Manin obstruction to the Hasse principle is known to be the only obstruction
for many types of homogeneous spaces of linear algebraic groups (Sansuc,
Borovoi, building on the results of Kneser, Harder, Chernousov; the first
counter-example to the Hasse principle for a principal homogeneous space
of a semisimple group was found by Serre). This is one possible generalization of the Minkowski–Hasse theorem for quadrics. In another direction,
computer calculations along with various partial or conditional results provide ample evidence that the Manin obstruction to the Hasse principle is
the only obstruction for varieties given by one cubic or two quadratic equations. Most of these results were obtained using a generalization of the
classical descent on cubic curves by Colliot-Thélène and Sansuc, who were

motivated by some pioneering computations of Swinnerton-Dyer. They introduced an important notion of universal torsors which encompasses descent on projective varieties with finitely generated geometric Picard group.

What happens next? Can we still hope for the existence of an algorithm when the Manin obstruction fails to be the only obstruction to the Hasse principle? The first counter-example to the Hasse principle not accounted for by the Manin obstruction was recently found by the author. It is a bielliptic surface over $\mathbf{Q}$, and can be given by two homogeneous quartic equations in five variables. Here again, but for a different reason, we believe that for such systems of equations a solubility algorithm should exist. To explain why we need to introduce *torsors*.

Suppose that we have a variety X over a field k, and an algebraic k-group G which can be finite or not. An X-torsor under G is a pair consisting of a morphism $Y \to X$ and an action of G on Y preserving f, which is locally (in an appropriate topology) a direct product. From the point of view of the geometric invariant theory a torsor is simply a variety Y equipped with a free (in the scheme-theoretic sense) action of G such that X is the quotient of Y by this action. One extreme case is when X is a point, $X = Spec(k)$, and G is a connected algebraic group, say an abelian variety or a linear group. Then k-torsors Y under G are also known under the classical name of *principal homogeneous spaces* of G. Another classical situation is when a finite group G acts freely on Y; then Y is an unramified Galois covering of $X = Y/G$ with group G.

Suppose that Y is a principal homogeneous space of an elliptic curve E defined over k, G is a finite subgroup of E, and $X = Y/G$. Assume that X has rational points over all completions of k. Then the classical procedure of *descent* sometimes tells us that X contains no k-rational point[1].

The descent method can be described for general torsors. One associates to an X-torsor Y under G and a k-torsor P under G a *twist* of Y by P, denoted by $_PY$. In practice a k-torsor P under G is represented by a 1-cocycle of the Galois group of k with coefficients in G. The twist $_PY$ is an X-torsor under a certain twisted form of G, which is G itself when G is abelian. Suppose now that X has points everywhere locally, that is, in all completions of k. Given a torsor $Y \to X$ the following descent method

---

[1] There are other applications of descent: when $Y = E$ the descent can be used to bound the number of generators of the group of k-rational points of E. In the weak Mordell–Weil theorem this is used to prove that the quotient of the group of rational points $E(k)$ of an elliptic curve E over a number field k by $mE(k)$ is finite for any $m \in \mathbf{Z}$.

can sometimes tell us that X has no k-rational point. It is possible to determine a finite set of k-torsors P under G such that if ${}_PY$ has points everywhere locally, then P is isomorphic to a k-torsor in this set. For example, if G is finite, and $Y \to X$ is an isogeny of elliptic curves, then this set is none other than the Selmer group associated to this isogeny. If no twist ${}_PY$ has points everywhere locally, this is an obstruction to the Hasse principle. Indeed, if X has a k-point M, then twisting Y by the k-torsor given by the fibre Y_M we arrive at a twisted torsor with a k-point in the fibre over M, hence in particular with points everywhere locally. We call this obstruction *the descent obstruction*. It was introduced by Colliot-Thélène and Sansuc in the case when G is a torus, but was already widely used in the classical descent theory on elliptic curves, where G is finite abelian. This was extended to the case of non-commutative groups by D. Harari and the author.

The point is that, at least in principle, this 'neo-classical' descent obstruction is computable in finitely many elementary operations: we only have to check the local solubility of a finite family of torsors in finitely many critical places. Another important feature is that the Manin obstruction can be recovered as a particular case of the descent obstruction. There are natural torsors attached to the varieties we discussed above: for cubic curves G is finite and abelian, for cubic surfaces G is a torus, and for bielliptic surfaces G is a finite nilpotent group of class 2. The counter-example to the Hasse principle not explained by the Manin obstruction, which we mentioned above, is explained by the descent obstruction related to a torsor under such a finite nilpotent group.

This book consists of two parts. The first part concerns the theory of torsors in general, and the second part is devoted to applications of torsors to the arithmetic of varieties over number fields. The general theory is given in Chapter 2. It is not surprising that a satisfactory classification is available in the abelian case, more precisely, for torsors under groups of multiplicative type (these are extensions of finite abelian group schemes by tori). We illustrate the general concepts by considering in detail some important examples of torsors in Chapter 3. These are quotients by 'generically free' torus actions and homogeneous spaces of algebraic groups. We consider the quotient of the Grassmannian $G(2,5)$ by the action of a maximal torus of $PGL(5)$, explicit 2- and 4-coverings of elliptic curves, the universal covering of a semisimple group. Chapter 4 deals with specific properties of torsors under groups of multiplicative type. The main result here is a description of certain torsors by explicit equations, which is a general set-up of 'explicit descent'. This is illustrated on the example of some natural torsors on hyperelliptic curves and conic bundle surfaces. In Chapter 5 we define the number-theoretic obstructions to the Hasse principle and to various approximation properties. The main theorem in Chapter 6 describes an important particular case of the Manin obstruction (the so called 'algebraic' Manin obstruction) in terms of the descent obstruction associated with torsors under groups of multiplicative type.

For this book we selected three recent applications of these general methods. In Chapter 7 we prove that the Manin obstruction to the Hasse principle is the only obstruction on some surfaces fibred into conics over the projective line, including some cases with six degenerate fibres. In Chapter 8 we study a surface of a different kind, which is a counter-example to the Hasse principle not explained by the Manin obstruction but explained by the non-abelian descent obstruction. In Chapter 9 using the language of gerbs we prove that the homogeneous spaces of (simply connected) semi-simple groups with connected stabilizers also have the property that the Manin obstruction is the only obstruction to the Hasse principle. These last three chapters illustrate different methods one uses, besides the theory of torsors, in studying the arithmetic of rational points in three main cases where some understanding of the Hasse principle is available. Roughly speaking, these are varieties which are either (1) close to rational varieties (quadrics, families or intersections of such, cubics, etc.), or (2) close to abelian varieties, or (3) close to affine algebraic groups (with methods using the group structure).

Note that the descent method has other important applications: the study of the R-equivalence classes of rational points, of the group of 0-cycles; descent provides an obstruction to weak approximation, etc. The scope of this book did not permit us to treat these and other important subjects closely related to the descent techniques. Some of this material is covered in the survey articles [C86], [C87], [C92], [C95], [C98], [MT], [Sansuc 82], [Sansuc 87], [SD96]. Another noticeable omission is the theory of the Brauer group.

Acknowledgements. The subject of this book owes its existence to the pioneering ideas of Yuri Ivanovich Manin and Sir Peter Swinnerton-Dyer, and the conceptual framework laid out by Jean-Louis Colliot-Thélène and Jean-Jacques Sansuc. I would like to thank Jean-Louis Colliot-Thélène for his generosity and openness. I am grateful to David Harari for permission to include here parts of our yet unpublished work, and for his comments. Thanks are due to M. Borovoi, B. Kunyavskiĭ and T. Szamuely for their comments, and to S. Siksek for his help with 4-descent. A large part of this book was written at I.H.E.S. (Bures-sur-Yvette) whose hospitality is gratefully acknowledged. I thank my wife Anna for suggesting the epigraph.

Part one

TORSORS

Notation and conventions. In this part k is a field of characteristic 0 (unless otherwise stated, often a less restrictive condition would be enough); $\overline{k}$ denotes a separable closure of k, $\Gamma_k = Gal(\overline{k}/k)$. By a *variety* over k we mean a separable scheme of finite type $p : X \to Spec(k)$. If X is a k-variety we write the action of Γ_k on the set of $\overline{k}$-points $X(\overline{k})$ by $(g, x) \mapsto {}^g x$, where $g \in \Gamma_k, x \in X(\overline{k})$. For a k-variety X we write $\overline{X} = X \times_k \overline{k}$.

We employ the notation $k[X] = \Gamma(X, \mathcal{O})$, $k[X]^* = \Gamma(X, \mathcal{O}^*)$. The set of Z-morphisms of Z-schemes $Y \to Y'$ is denoted by $Mor_Z(Y, Y')$, in particular, $Mor_k(X, \mathbf{A}_k^1) = k[X]$.

Let $\mathbf{G}_m = \mathbf{G}_{m,\mathbf{Z}} = Spec(\mathbf{Z}[t, t^{-1}])$, this is a group scheme over $Spec(\mathbf{Z})$, which as a scheme is isomorphic to $\mathbf{A}_{\mathbf{Z}}^1 \setminus \{0\}$. We have $Mor_k(X, \mathbf{G}_{m,k}) = k[X]^*$.

Unless otherwise stated all cohomology and sheaves are understood in the sense of the étale topology. The sheaf $\mathbf{G}_m$ over a scheme X is defined by $\mathbf{G}_m(U) = \Gamma(U, \mathcal{O}^*) = Mor_X(U, \mathbf{G}_{m,X})$ for $U \to X$ étale. A group scheme G over Z is called a Z-*group of multiplicative type* if it is locally (for the étale topology) a group subscheme of $\mathbf{G}_{m,Z}^n$ for some n. Torsors under groups of multiplicative type will be sometimes referred to as *abelian torsors*. A group scheme G over Z is called a Z-*torus* if it is locally isomorphic to $\mathbf{G}_{m,Z}^n$ for some n.

The *Brauer group* $Br(X)$ throughout this book is understood as the cohomological Brauer–Grothendieck group $H^2(X, \mathbf{G}_m)$. Let $Br_0(X)$ (resp. $Br_1(X)$) be the image of the natural map $p^* : Br(k) \to Br(X)$ (resp. the kernel of the natural map $Br(X) \to Br(\overline{X})^{\Gamma_k}$).

By $M[n]$ we denote the n-torsion subgroup of an abelian group M. If M is a discrete Γ_k-module we shall write $H^i(k, M)$ for the Galois cohomology groups $H^i(\Gamma_k, M)$. If R is a commutative ring, and $\mathcal{F}$ is a sheaf over $Spec(R)$ we write $H^i(R, \mathcal{F})$ for $H^i(Spec(R), \mathcal{F})$.

The notation $Ext_k^n(\cdot, \cdot)$ (resp. $Ext_X^n(\cdot, \cdot)$, resp. $Ext_{k\text{-groups}}^n(\cdot, \cdot)$) stands for the group of n-fold extensions in the category of Γ_k-modules (resp. of sheaves over X, resp. of commutative algebraic k-groups). We write $\mathcal{H}om_X$ for the internal Hom of sheaves on X, and $\mathcal{H}om_k$ for the internal Hom of sheaves on $Spec(k)$.

An *algebraic k-group* is a k-group scheme which is a smooth variety over k (hence of finite type). Recall that any algebraic group is quasi-projective as an algebraic variety. An algebraic k-group S such that $\overline{S}$ is a group subscheme of $\mathbf{G}_{m,\overline{k}}^n$ for some n is a k-group of multiplicative type. A connected k-group of multiplicative type is a k-torus. The *module of characters* of S is $Hom_{k\text{-groups}}(S, \mathbf{G}_m)$; it is denoted by $\hat{S}$. Associating $\hat{S}$ to S defines an anti-equivalence of the category of k-groups of multiplicative type with the category of discrete Γ_k-modules which are of finite type as abelian groups. If $\tau : S_1 \to S_2$ is a homomorphism of k-groups of multiplicative type, then we denote by $\hat{\tau} : \hat{S}_2 \to \hat{S}_1$ the dual map of Γ_k-modules, and vice versa. A k-group S of multiplicative type is a torus if and only if $\hat{S}$ is torsion free. A k-torus is called *quasi-trivial* if its module of characters is a *permutation Γ_k-module*, which means by definition that it has a Γ_k-invariant basis.

2

Torsors: general theory

The aim of the first two sections of this chapter is to review the general theory of torsors with its useful tools, like twisting by Galois, *fppf* or étale descent, Čech cohomology, contracted product. We included only the concepts relevant for our purposes in this book, and the reader who needs to know more is referred to [Milne, EC], [BLR] and [Giraud]. The reader interested in applications of torsors can skip most of Section 2.2 in the first reading.

Section 2.3 contains the theory of torsors under groups of multiplicative type which is due to Colliot-Thélène and Sansuc. This theory is much more precise than the general case. Here belongs the notion of *the type* of the torsor, and that of *universal torsors*, which leads to the so called *elementary obstruction* to the existence of rational points (defined over arbitrary fields).

In the final section we relate this elementary obstruction to another one, earlier defined by Grothendieck. It is the class of the well known natural extension of the absolute Galois group of k by the (geometric) fundamental group of $\overline{X}$, given by the algebraic fundamental group of X. The significance of this obstruction for the study of rational points remains to be explored.

2.1 Torsors over a field

Let us start by recalling the definition of torsors under an algebraic (possibly non-abelian) group defined over a field k. An unsurpassed exposition of the first non-abelian cohomology set is found in [Serre, CG], I.5, and the reader is expected to be to a certain extent familiar with it. Torsors (in other terminology, principal homogeneous spaces) are discussed in [Serre, GA], V.4.21 and [Serre, CG], III.1.

Let G be an algebraic group defined over k. Its left action on itself

11

is $(s, x) \mapsto sx$, and the right action is $(s, x) \mapsto xs$. These actions are compatible with the action of Γ_k: for $s_1, s_2 \in G(\overline{k})$, $g \in \Gamma_k$ we have $^g(s_1 s_2) = {}^g s_1 \cdot {}^g s_2$.

Unless otherwise stated, a torsor will always mean a right torsor. We leave it to the reader to modify the following definition for left torsors.

Definition 2.1.1 *A k-torsor under G is a non-empty k-variety X equipped with a right action of G which turns $X(\overline{k})$ into a principal homogeneous space under $G(\overline{k})$: for any $x_1, x_2 \in X(\overline{k})$ there exists a unique $s \in G(\overline{k})$ such that $x_2 = x_1 s$. In other words, a k-torsor is a $(\overline{k}/k)$-form of the variety G with the right action of G, more precisely, this is a pair consisting of a k-variety and an action of G, which over $\overline{k}$ becomes isomorphic to G with its right action on itself.*

An isomorphism of $\overline{G}$-torsors $\overline{G} \to \overline{X}$ is uniquely determined by the image of the neutral element.

An important fact is that k-torsors are quasi-projective (in particular, the same is true for G itself). This result also holds with $Spec(k)$ replaced by the spectrum of a Dedekind ring (see [BLR], Thm. 6.4.1).

Twisting by Galois descent.

Let F be a quasi-projective k-variety endowed with an action of G. Let $\sigma : \Gamma_k \to G(\overline{k})$ be a continuous 1-cocycle, Γ_k being endowed with its natural profinite topology, and $G(\overline{k})$ with discrete topology. This is a continuous map which is a crossed homomorphism, that is, a map satisfying $\sigma_{g_1 g_2} = \sigma_{g_1} \cdot {}^{g_1}\sigma_{g_2}$. The 'twist' of F by σ, denoted by F^σ, is defined as the quotient of $\overline{F}$ by the twisted action of Γ_k given by $(g, s) \mapsto \sigma_g \cdot {}^g s$. This quotient exists by Weil's theorem on descent of the base field ([Serre, GA], V.4, Cor. 2; see [BLR], Chapter 6, for the general exposition of Grothendieck's descent theory). Replacing σ by a cohomologous cocycle $\sigma'_g = c^{-1}\sigma_g {}^g c$ (for some $c \in G(\overline{k})$) gives rise to an isomorphic variety. The isomorphism depends on the choice of c and so is not canonical.

Example 1: inner forms.

Applying this to $F = G$ acting on itself by conjugations we arrive at the notion of the inner form G^σ of G. This is an algebraic group over k which is a $(\overline{k}/k)$-form of G as an algebraic group. In general not every $(\overline{k}/k)$-form of the algebraic k-group G can be obtained in this way. Let $Inn(G) \subset Aut(G)$ be the subgroup of inner automorphisms of G. We have $Inn(G) = G/Z(G)$, where $Z(G)$ is the centre of G. Inner forms are classified by $H^1(k, Inn(G))$, as opposed to arbitrary forms of G classified by $H^1(k, Aut(G))$. Note that if G is commutative, then G^σ and G are canonically isomorphic.

Example 2: torsors.

The group G acting on itself on the left is the automorphism group of the pair $(G,$ the right action of G on $G)$. Thus the corresponding twisted variety is equipped with the right action of G making it into a right torsor under G. To avoid confusion with inner forms, we shall denote this torsor by X^σ.

Conversely, any k-torsor X under G can be obtained in this way: choose a $\overline{k}$-point $\overline{x}_0 \in X(\overline{k})$; then for any $g \in \Gamma_k$ there is a unique element $\sigma_g \in G(\overline{k})$ such that ${}^g\overline{x}_0 = \overline{x}_0\sigma_g$. The map $g \mapsto \sigma_g$ is a continuous 1-cocycle. (It is trivial on the subgroup $\Gamma_K \subset \Gamma_k$ if $\overline{x}_0$ comes from a K-point of X.) One sees immediately that different $\overline{k}$-points $\overline{x}_0$ lead to cohomologous cocycles. These two constructions being inverse to each other ($\overline{x}_0$ corresponds to the neutral element of $G(\overline{k})$), we obtain a bijection

$$\begin{array}{c} k\text{-torsors under } G \\ \text{up to isomorphism} \end{array} \longleftrightarrow \text{ the pointed set } H^1(k, G) = H^1(\Gamma_k, G(\overline{k})).$$

$$(2.1)$$

The distinguished point represents the class of the trivial torsor, that is, of G with its right action on itself.

It is less obvious but easy to check directly that in addition to the structure of a right torsor under G, X^σ has the structure of the left torsor under G^σ, that is, G^σ acts on X^σ on the left compatibly with the action of Γ_k, making X^σ into a left torsor under G^σ. (This could be rephrased by saying that X^σ is a G^σ-G-bitorsor; see [Serre, CG], I.5.34.)

2.2 Torsors and Čech cohomology

We now review the theory of torsors over a general base scheme (see [Milne, EC], III.4 for more details; see also [Giraud], III). Till the end of this section, unless otherwise stated, the topology is the flat topology, that is, *fppf*-topology. This traditional French abbreviation means *faithfully flat and of finite presentation*.

Definition 2.2.1 *Let X be a scheme. An X-**torsor** under an X-group scheme G is defined as a scheme Y/X equipped with an action of G compatible with the projection to X and satisfying the following equivalent properties:*

(i) the morphism $Y \to X$ is fppf, and the map $Y \times_X G \to Y \times_X Y$ given by $(y, s) \mapsto (y, ys)$ is an isomorphism;

(ii) there exists a covering $\mathcal{U} = (U_i \to X)$ in the flat topology such that for any i the pair $(Y_{U_i} = Y \times_X U_i$, the action of $G_{U_i} = G \times_X U_i)$ is isomorphic to the pair $(G_{U_i}$, the right action of G_{U_i} on itself).

Proof of the equivalence of (i) and (ii). ([Milne, EC], I.4.1) The flat topology is the most natural topology for the study of torsors since $Y \to X$ is itself a covering. Thus the implication $(i) \Rightarrow (ii)$ is obvious yet meaningful: there is a 'canonical' covering trivializing $Y \to X$, namely, $Y \to X$ itself. To prove the converse we let U be the disjoint union of the U_i's. The morphism $U \to X$ is faithfully flat and locally of finite type. The pull-back of Y to U is isomorphic to the pull-back of G to U, $Y_U = G_U$. Hence $Y_U \to U$ has all the properties mentioned in (i). Grothendieck's theory of descent with respect to morphisms which are faithfully flat and locally of finite type (see [Milne, EC], 1.2.24) implies that $Y \to X$ also has these properties. QED

The torsor $G \to X$ with the right action of G is called trivial. A torsor $Y \to X$ is trivial if and only if the morphism $Y \to X$ has a section. Indeed, a section y defines an isomorphism $G \to Y$ given by $s \mapsto ys$.

An important property of torsors is that the category of X-torsors under G is a groupoid, or, in other words, any morphism $Y_1 \to Y_2$ compatible with canonical projections to X and the action of G is an isomorphism. Indeed, any morphism $\phi : G \to G$ compatible with the canonical projection to X sends the unit section $X \to G$ to some section $s_0 : X \to G$. Compatibility with the right action of G implies that ϕ is given by $s \mapsto s_0 s$ which is obviously an isomorphism. Now our statement follows by the *fppf* descent with respect to a covering which trivializes both Y_1 and Y_2.

If in Definition 2.2.1 we replace Y by a sheaf of sets on X in the *fppf*-topology with a right action of a sheaf of groups satisfying *(ii)*, we arrive at the notion of a sheaf of torsors. In general, not every sheaf of torsors under an X-group scheme G is represented by a scheme. However, this is so, if, for instance, G is affine over X ([Milne, EC], III.4.3 (a)). It is a theorem of Raynaud that the same is true if G is regular, and G/X is smooth and proper with geometrically connected fibres (cf. *loc. cit.*). In the sequel G will always fall within one of these two classes.

The isomorphism classes of torsors are naturally described by the elements of the first non-abelian Čech cohomology set. Before defining it let us recall the usual definition of the Čech cohomology with coefficients in a presheaf of abelian groups $\mathcal{P}$.

Abelian Čech cohomology.

Let $\mathcal{U} = (U_i \to X)_{i \in I}$ be a covering in some fixed (Grothendieck) topology, usually *fppf* or étale. We write $U_{ij} = U_i \times_X U_j$, $U_{ijk} = U_i \times_X U_j \times_X U_k$, and so on. The canonical projections

$$p_{i_j} : U_{i_0, \ldots, i_n} \longrightarrow U_{i_0, \ldots, \hat{i}_j, \ldots, i_n}$$

induce the maps

$$p_{i_j}^* : \mathcal{P}(U_{i_0, \ldots, \hat{i}_j, \ldots, i_n}) \longrightarrow \mathcal{P}(U_{i_0, \ldots, i_n}).$$

The Čech complex is

$$\check{C}^n(\mathcal{U}, \mathcal{P}) = \prod_{i \in I^{n+1}} \mathcal{P}(U_{i_0, \ldots, i_n})$$

with the differential

$$d_{i_0, \ldots, i_n}^n = \sum_{j=0}^{n+1} (-1)^j p_{i_j}^*.$$

The Čech groups $\check{H}^n(\mathcal{U}/X, \mathcal{P})$ are the cohomology groups of this complex, and $\check{H}^n(X, \mathcal{P})$ can be defined by passing to the inductive limit for all coverings (see [Milne, EC], III.2). Note that there is a natural map $\mathcal{P}(X) \to \check{H}^0(\mathcal{U}/X, \mathcal{P})$ which is an isomorphism when $\mathcal{P}$ is a sheaf. There is the Čech spectral sequence $\check{H}^p(\mathcal{U}/X, \mathcal{H}^q(\mathcal{P})) \Rightarrow H^{p+q}(X, \mathcal{P})$, where $\mathcal{H}^q(\mathcal{P})$ is the presheaf $U \longmapsto H^q(U, \mathcal{P})$ ([Milne, EC], III.2.7).

If a morphism $Y \to X$ is a covering this gives a spectral sequence

$$\check{H}^p(Y/X, \mathcal{H}^q(\mathcal{P})) \Rightarrow H^{p+q}(X, \mathcal{P}). \tag{2.2}$$

The corresponding exact sequence of low degree terms begins as follows (we assume here that $\mathcal{P}$ is a sheaf):

$$\begin{aligned} 0 &\to \check{H}^1(Y/X, \mathcal{P}) \to H^1(X, \mathcal{P}) \to \check{H}^0(Y/X, \mathcal{H}^1(\mathcal{P})) \\ &\to \check{H}^2(Y/X, \mathcal{P}) \to H^2(X, \mathcal{P}). \end{aligned} \tag{2.3}$$

This sequence is very useful. Let us give two examples of X-torsors when (2.2) and (2.3) have explicit descriptions.

Example 1: Hochschild–Serre spectral sequence.

We follow ([Milne, EC], Example III.2.6). Let F be a finite group. A finite étale Galois covering Y/X with Galois group F is an X-torsor under an X-group scheme F_X which as an X-scheme is the disjoint union of $|F|$ copies of X with the group structure inherited from that of F.

For any sheaf $\mathcal{P}$ (actually, any presheaf such that $\mathcal{P}(X_1 \cup X_2) = \mathcal{P}(X_1) \times \mathcal{P}(X_2)$ if X_1 and X_2 are disjoint) one has $\mathcal{P}(Y \times F^n) = Hom_{sets}(F^n, \mathcal{P}(Y))$. A direct verification then shows that the Čech complex $\check{C}^{\cdot}(Y/X, \mathcal{P})$ is isomorphic to the complex of non-homogeneous cochains of the group F with coefficients in $\mathcal{P}(Y)$. Thus the Čech cohomology groups of the canonical covering are computed in terms of group cohomology: $\check{H}^i(Y/X, \mathcal{P}) = H^i(F, \mathcal{P}(Y))$.

Suppose now that our topology is flat or étale. Then the Čech spectral sequence associated to the canonical covering is the Hochschild–Serre spectral sequence ([Milne, EC], III.2.20, 2.21):

$$H^p(F, H^q(Y, \mathcal{P})) \Rightarrow H^{p+q}(X, \mathcal{P}).$$

Passing to the limit one extends this to profinite Galois coverings (*ibidem*).

Example 2.

Let us compute $\check{H}^1(Y/X, \mathbf{G}_m)$ when X and Y are k-varieties such that $f : Y \to X$ is a torsor under an algebraic k-group G in the flat or étale topology. By definition, $\check{H}^1(Y/X, \mathbf{G}_m)$ is the middle cohomology group of the complex

$$H^0(Y, \mathbf{G}_m) \to H^0(Y \times_X Y, \mathbf{G}_m) \to H^0(Y \times_X Y \times_X Y, \mathbf{G}_m),$$

where the first arrow written additively is $p_1^* - p_0^*$, and the second one is $p_{01}^* + p_{12}^* - p_{02}^*$. Applying the isomorphisms $Y \times_X Y = Y \times_k G$ and $Y \times_X Y \times_X Y = Y \times_k G \times_k G$ we can rewrite this complex as

$$H^0(Y, \mathbf{G}_m) \to H^0(Y \times_k G, \mathbf{G}_m) \to H^0(Y \times_k G \times_k G, \mathbf{G}_m).$$

The middle cohomology group is the group of invertible regular functions h on $Y \times_k G$ such that $h(y, s)h(ys, s') = h(y, ss')$ modulo functions of the form $g(ys)g(y)^{-1}$, where g is an invertible function on Y. The first relation is just the cocycle condition; the functions $g(ys)g(y)^{-1}$, where g is an invertible function on Y, are coboundaries. Note that the condition on h is satisfied by a function $Y \times_k G \to \mathbf{G}_m$ coming from a character of G.

Variant 1.

If Y satisfies $H^0(\overline{Y}, \mathbf{G}_m) = \overline{k}^*$, then $H^0(Y \times_k G, \mathbf{G}_m)$ is isomorphic to $H^0(G, \mathbf{G}_m)$, and hence any function satisfying the cocycle condition comes from a character of G. The only coboundary is the constant function 1. Thus $\check{H}^1(Y/X, \mathbf{G}_m) = \hat{G}(k)$, and (2.3) gives rise to the exact sequence

$$1 \to \hat{G}(k) \to Pic(X) \xrightarrow{f^*} Pic(Y). \tag{2.4}$$

This sequence is often applied to quotients of freely acting finite groups.

This sequence is functorial with respect to field extensions. Writing it for $\overline{k}$ instead of k we obtain an exact sequence of Γ_k-modules

$$1 \to \hat{G} \to Pic(\overline{X}) \xrightarrow{f^*} Pic(\overline{Y}). \qquad (2.5)$$

Variant 2.

Assume that X, Y and G are geometrically integral, and G is a (connected) linear algebraic k-group. A well known lemma of Rosenlicht is the statement that *any invertible function on G is a product of a constant and a character $G \to \mathbf{G}_m$* (see, e.g., [Sansuc 81]). Since Y is geometrically connected, any invertible function on $Y \times_k G$ must be of the form $\phi(y)\chi(s)$, where $\phi \in H^0(Y, \mathbf{G}_m)$ and $\chi \in \hat{G}(k)$. The cocycle condition then implies that ϕ is identically 1. We thus obtain a surjection $\hat{G}(k) \to \check{H}^1(Y/X, \mathbf{G}_m)$. Any element in its kernel comes from $H^0(Y, \mathbf{G}_m)$ via the map sending g to the character of G given by $g(ys)g(y)^{-1}$ (this makes sense because of the previous argument based on Rosenlicht's lemma). Since Y is an X-torsor, the G-invariant invertible functions on Y come from X. Summing up, we obtain the exact sequence of abelian groups

$$1 \to H^0(X, \mathbf{G}_m) \to H^0(Y, \mathbf{G}_m) \to \hat{G}(k) \to Pic(X) \to Pic(Y), \qquad (2.6)$$

which can also be written as

$$1 \to H^0(X, \mathbf{G}_m)/k^* \to H^0(Y, \mathbf{G}_m)/k^* \to \hat{G}(k) \to Pic(X) \to Pic(Y). \qquad (2.7)$$

This sequence is also functorial with respect to field extensions. We thus obtain an exact sequence of Γ_k-modules

$$1 \to \overline{k}[X]^*/\overline{k}^* \to \overline{k}[Y]^*/\overline{k}^* \to \hat{G} \to Pic(\overline{X}) \to Pic(\overline{Y}). \qquad (2.8)$$

If moreover G is assumed to be reductive if k is not perfect, then using Rosenlicht's lemma one can extend these sequences to the right, and obtain Sansuc's exact sequences ([Sansuc 81], 6.10).

When G is an *algebraic torus* Colliot-Thélène and Sansuc ([CS87a], 2.1.1) show that the last arrow in (2.8) is *surjective*. We are not going to use this result in what follows, so we only briefly sketch the proof. If $f : Y \to X$ is a torsor under a torus T, then the relative version of Rosenlicht's lemma ([CS87a], 1.4.2) states that there is a natural exact sequence of étale sheaves on X:

$$1 \to \mathbf{G}_m \to f_*\mathbf{G}_m \to \hat{T} \to 0.$$

Moreover, the class of this extension, properly interpreted (see Lemma 2.3.7 below), coincides up to sign with the class of the torsor Y/X in $H^1(X,T)$ ([CS87a], 1.4.3). Let K be any extension of k contained in $\overline{k}$. The long exact sequence of étale cohomology obtained from the above extension is

$$1 \to K[X]^* \to K[Y]^* \to \hat{T}^{Gal(\overline{k}/K)} \to Pic(X_K) \to Pic(Y_K) \to 0. \quad (2.9)$$

When $K = \overline{k}$, it coincides with (2.8) up to inverting the signs ([CS87a], 2.1.1).

Non-abelian Čech cohomology.

Let us now recall the definition of the first non-abelian Čech cohomology set with coefficients in a sheaf of groups $\mathcal{G}$. When $\mathcal{G}$ is represented by an X-group scheme G we shall write G instead of $\mathcal{G}$. A 1-cocycle with respect to the covering $\mathcal{U} = (U_i \to X)$ is a family $s_{ij} \in \mathcal{G}(U_{ij})$ for all i and j such that after the restriction to U_{ijk} we have $s_{ij}s_{jk} = s_{ik}$. The cocycles s and s' are cohomologous if there exist elements $h_i \in \mathcal{G}(U_i)$ such that after the restriction to U_{ij} we have $s'_{ij} = h_i s_{ij} h_j^{-1}$. The pointed set of cohomology classes is denoted by $\check{H}^1(\mathcal{U}/X, \mathcal{G})$ (the distinguished element is represented by $s_{ij} = 1$). Passing to the inductive limit for all coverings we obtain the set $\check{H}^1(X, \mathcal{G})$.

Let $\mathcal{Y} \to X$ be a sheaf of torsors under $\mathcal{G}$ trivialized on a covering $\mathcal{U} = (U_i \to X)$. Choosing local sections $y_i \in \mathcal{Y}(U_i)$ there exists a unique $s_{ij} \in \mathcal{G}(U_{ij})$ such that $y_i s_{ij} = y_j$ on $\mathcal{G}(U_{ij})$. The family $\{s_{ij}\}$ is a 1-cocycle with coefficients in $\mathcal{G}$. This associates to a sheaf of torsors over X under $\mathcal{G}$ trivialized by $\mathcal{U}$ a class in $\check{H}^1(\mathcal{U}/X, \mathcal{G})$. The distinguished element of $\check{H}^1(\mathcal{U}/X, \mathcal{G})$ corresponds to the sheaf of trivial torsors $\mathcal{G}$. This defines a bijection, more precisely, an isomorphism of pointed sets ([Milne, EC], III.4.6):

$$\begin{array}{c} \text{sheaves of torsors over } X \text{ under } \mathcal{G} \\ \text{trivialized on } \mathcal{U}, \text{ up to isomorphism} \end{array} \longleftrightarrow \text{ the pointed set } \check{H}^1(\mathcal{U}/X, \mathcal{G}).$$

Passing to the inductive limit we obtain a bijection

$$\begin{array}{c} \text{sheaves of torsors over } X \text{ under } \mathcal{G} \\ \text{up to isomorphism} \end{array} \longleftrightarrow \text{ the pointed set } \check{H}^1(X, \mathcal{G}).$$

If G is such that every X-sheaf of torsors under G is represented by an X-scheme, e.g., if G/X is affine, we have a bijection

$$\begin{array}{c} X\text{-torsors under } G \\ \text{up to isomorphism} \end{array} \longleftrightarrow \text{ the pointed set } \check{H}^1(X, G). \quad (2.10)$$

If G is commutative, we can replace the Čech cohomology group by the flat cohomology group:

X-torsors under G up to isomorphism $\longleftrightarrow$ the group $H^1(X, G)$. (2.11)

Indeed, $H^1(X, F)$ of a sheaf of abelian groups F can always be computed as $\check{H}^1(X, F)$ ([Milne, EC], III.2.10).

If G/X is smooth (for instance, if X is a variety over a field k, and G comes from an algebraic k-group), the flat topology can be replaced by the étale topology ([Milne, EC], III.4). Thus when G is commutative, we have a full analogy with (2.1): X-torsors under G are classified by the elements of the group $H^1_{\acute{e}t}(X, G)$ ([Milne, EC], I.4.7).

The cohomology class of a torsor $Y \to X$ in the relevant cohomology set (or group) is denoted by $[Y]$.

Let us go back to the general case of a sheaf of groups $\mathcal{G}$ over X. As is usual with non-abelian cohomology, one can define a non-abelian analogue of the first three terms of (2.3) by *ad hoc* considerations. Let $\check{\mathcal{H}}^0(\mathcal{G})$ (resp. $\check{\mathcal{H}}^1(\mathcal{G})$) be the obvious presheaf of groups (resp. of pointed sets). For any sheaf of sets we have $\mathcal{G} = \check{\mathcal{H}}^0(\mathcal{G})$. Then there is an exact sequence of pointed sets

$$1 \to \check{H}^1(\mathcal{U}/X, \mathcal{G}) \to \check{H}^1(X, \mathcal{G}) \to \check{H}^0(\mathcal{U}/X, \check{\mathcal{H}}^1(\mathcal{G})).$$

The last arrow is given by the collection of restrictions from X to U_i, and $\check{H}^1(\mathcal{U}/X, \check{\mathcal{H}}^0(\mathcal{G}))$ parametrizes the classes of cocycles trivialized on $\mathcal{U}$.

Lemma 2.2.2 *Let G and G' be algebraic groups over k, and X and Y be k-varieties such that $Y \to X$ is a torsor under G. There is an exact sequence of pointed sets*

$$1 \to \check{H}^1(Y/X, G') \to \check{H}^1(X, G') \to \check{H}^0(Y/X, \check{\mathcal{H}}^1(G')).$$

The pointed set $\check{H}^1(Y/X, G')$ can be interpreted as the set of equivalence classes of morphisms $f : Y \times_k G \to G'$ satisfying the cocycle condition $f(y, s)f(ys, s') = f(y, ss')$; f is equivalent to f' if

$$f'(y, s) = g(y)f(y, s)g(ys)^{-1}$$

for a morphism $g : Y \to G'$. If $G = G'$, then the class of the torsor $Y \to X$ in $\check{H}^1(Y/X, G)$ is given by the second projection $Y \times_k G \to G$.

Proof. All statements except the last one are straightforward. The last statement is verified directly from the definitions. Indeed, the torsor $Y \to X$ is trivialized by the covering $Y \to X$, and the map $Y \times_X Y \to Y$ has a section given by the diagonal. Then the cocycle of $Y \to X$ in $H^0(Y \times_X Y, G)$ becomes the second projection after the isomorphism $H^0(Y \times_X Y, G) = H^0(Y \times_k G, G)$. QED

Contracted product. Twisting by fppf descent.

This important construction is crucial for the applications of torsors.

Lemma 2.2.3 *Let P be a right X-torsor under an X-group scheme G, and F be an affine X-scheme equipped with a left action of G (compatible with the projection to X). Then the quotient of $P \times_X F$ by the action of G given by $(p, f) \mapsto (ps^{-1}, sf)$ exists as an affine X-scheme (in other words, there exists a morphism of X-schemes $P \times_X F \to Y$ whose fibres are orbits of G).*

The proof of the lemma, which is a simple application of the *fppf* descent, will be given a little later.

The quotient whose existence is stated in the lemma is called *the contracted product* of P and F with respect to G, and is denoted by $P \times_X^G F$, or simply by $P \times^G F$. (All our fibre products are taken over X, so we shall sometimes omit the subscript X.) It is also called *the twist of F by P*, and is denoted by $_P F$. Note that P has the structure of a left X-torsor under $_P G$, so that $_P G$ acts on $_P F$ on the left.

Example 1: inner forms.

Let $F = G$ be acting on itself by conjugations. The contracted product is an X-group scheme $_P G$, which locally in the *fppf* topology is isomorphic to G. If $X = Spec(k)$, then $_P G$ is the inner form G^σ of G, where P is the k-torsor defined by $\sigma \in Z^1(\Gamma_k, G)$ (see Section 2.1).

Example 2: twisting an X-torsor.

Let $F = Q$ be a left X-torsor under G. Then it is also a right X-torsor under an appropriate inner form of G. (More precisely, let Q' be the *inverse* torsor of Q: as an X-scheme it is isomorphic to Q, and it is a right torsor under G with respect to the action $x's := s^{-1}x'$. Like any right torsor under G, Q' is also a left X-torsor under $G' := {}_{Q'}G$. Then Q is equipped with the structure of a right X-torsor under G' with respect to the action $xs' := s'^{-1}x$.) Summing up, the contracted product $P \times^G Q$ is a right X-torsor under G', and a left X-torsor under $_P G$. It is denoted by $P \circ Q$. For example we have $Q' \circ Q = G$. In other terms, the diagonal image of Q in $Q' \times_X Q$ is an orbit of G, leading to a section of the quotient X-scheme.

The operation $P \mapsto P \circ Q$ defines a bijection of sets $\check{H}^1(X, G) \to \check{H}^1(X, G')$, which sends the distinguished point to the class of Q. The inverse bijection is obtained by taking the contracted product with Q' with respect to G'. In the case when G is abelian, there is no difference between G and G', and the contracted product defines a group structure on $\check{H}^1(X, G)$, and the above bijection is just the translation by the class of Q.

When we have to twist a right X-torsor P under G with another right X-torsor E under G, we first consider the inverse E' which is a left torsor under G, and then form the contracted product $P \times^G E'$. This is just the quotient scheme of $P \times_X E$ by the simultaneous right action of G. In this case the twist $P \times^G E'$ is a right X-torsor under ${}_E G$, and is denoted by ${}_E P$. For example, ${}_P P$ is a trivial torsor under ${}_P G$. If G is abelian, the class of E' is the inverse of the class of E, hence in the group $H^1(X, G)$ we have a relation $[{}_E P] = [P] - [E]$.

We shall mostly deal with the case when X and P are varieties over k, G comes from an algebraic k-group, and $E = X \times_k Z$, where Z is a right k-torsor under G. Then ${}_E P$, also denoted by ${}_Z P$, can be obtained by Galois descent as in Section 2.1: take a cocycle $\sigma \in Z^1(\Gamma_k, G)$ defining Z, then consider the quotient $\overline{P}^\sigma$ of $\overline{P}$ by the corresponding twisted action of Γ_k, which is $(g, x) \mapsto {}^g x \sigma_g^{-1}$. Note that to use Galois descent we need the assumption that P is a quasi-projective k-variety.

Example 3: push-forward of torsors.

Let F be another X-group scheme together with a morphism $\alpha : G \to F$ of X-group schemes. Then F is equipped with a natural left action of G. The contracted product $P \times^G F$ is a right torsor under F. The resulting operation associating to a torsor under G a torsor under F is compatible with the push-forward map $\alpha_* : \check{H}^1(X, G) \to \check{H}^1(X, F)$ defined by applying α to the Čech cocycles. In the abelian case this is the usual map of cohomology groups.

Proof of Lemma 2.2.3. Let $p_1, p_2 : P \times P \to P$, and $p_{ji} : P \times P \times P \to P \times P$, for $j > i$, be the various projections. By Grothendieck's descent theory, to give an affine scheme Y over X amounts to the same thing as to give an affine scheme Y' over P together with an isomorphism $\varphi : p_1^* Y' \to p_2^* Y'$ satisfying the cocycle condition $p_{31}^*(\varphi) = p_{32}^*(\varphi) p_{21}^*(\varphi)$ (see [Milne, EC], I.2.23, [BLR], VI). By the property (i) in the definition of a torsor, the pull-back of $P \times F$ to P is isomorphic to $P \times G \times F$. Taking the contracted product $G \times^G F$ is the same thing as considering the map $G \times F \to F$ given by $(s, t) \mapsto st$. Hence $G \times^G F$ exists and is canonically isomorphic to F. Set $Y' = P \times F$, and let $\varphi : p_1^* Y' = P \times P \times F \to p_2^* Y' = P \times P \times F$ be given by $(x_1, x_2, t) \mapsto (x_1, x_2, s_{21}t)$, where $x_2 s_{21} = x_1$. This is clearly an isomorphism, and the cocycle condition obviously holds. This establishes the existence of Y. The map $P \times P \times F \simeq P \times G \times F \to P \times F = Y'$ (quotient by G acting as in the statement of the lemma) descends to $P \times F \to P \times^G F = Y$. QED

Partition of $X(k)$ defined by a torsor $Y \to X$.

Let $p : X \to Spec(k)$ be a variety over k, and G an algebraic k-group. If Z is a right k-torsor under G, let $_Z f : {}_Z Y \to X$ be the corresponding twisted right X-torsor under $_Z G$. (Recall that it exists provided Y is quasi-projective or G is affine.)

A torsor $f : Y \to X$ under G defines a map $\theta_Y : X(k) \to H^1(k, G)$ which associates to $P \in X(k)$ the class of the fibre Y_P in $H^1(k, G)$. This gives a partition of the set $X(k)$ into the subsets of points such that the corresponding fibres of f are isomorphic k-torsors under G,

$$X(k) = \bigcup_{a \in H^1(k,G)} \theta_Y^{-1}(a).$$

Using the twisting operation we can describe the partition of $X(k)$ defined by $Y \to X$ in a slightly different fashion. Let $P \in X(k)$, and choose Z isomorphic to the fibre Y_P of $Y \to X$ at P. Then we claim that the fibre of $_Z Y$ at P has k-rational points. Indeed, the diagonal $Z \hookrightarrow Z \times_k Z$ gives rise to the embedding

$$Spec(k) = Z/G \hookrightarrow (Z \times_k Z)/G = ({}_Z Y)_P.$$

(To put it differently, twisting commutes with base change, so we have $({}_Z Y)_P = {}_Z(Y_P) = {}_Z Z$, which is a trivial right torsor under $_Z G$.) The fibres of $_Z Y$ which contain k-points are located precisely over the points $P \in X(k)$ such that Y_P and Z are isomorphic right k-torsors under G. We summarize this by the formula

$$X(k) = \bigcup_Z {}_Z f\big({}_Z Y(k)\big). \tag{2.12}$$

Here Z runs over the set of k-torsors under G containing one representative from every isomorphism class.

The Leray spectral sequence.

We end this section by recalling a few basic facts about the Leray spectral sequence. Let $p : X \to Spec(k)$ be a k-variety, and S a sheaf of abelian groups on X. We have the Leray spectral sequence

$$H^p(k, R^q p_* S) \Rightarrow H^{p+q}(X, S). \tag{2.13}$$

The exact sequence of the first five low degree terms of (2.13) is

$$0 \longrightarrow H^1(k, p_*\mathcal{S}) \longrightarrow H^1(X, \mathcal{S}) \longrightarrow H^0(k, R^1p_*\mathcal{S})$$
$$\overset{\partial}{\longrightarrow} H^2(k, p_*\mathcal{S}) \longrightarrow H^2(X, \mathcal{S}). \tag{2.14}$$

Here the map $H^i(k, p_*\mathcal{S}) \to H^i(X, \mathcal{S})$, $i = 1, 2$, is the composition of the obvious map $H^i(k, p_*\mathcal{S}) \to H^i(X, p^*p_*\mathcal{S})$ with the map $H^i(X, p^*p_*\mathcal{S}) \to H^i(X, \mathcal{S})$ induced by the adjunction morphism $p^*p_*\mathcal{S} \to \mathcal{S}$. When S is an abelian k-group, we shall write S for the sheaf of abelian groups on k or X represented by S. Note that the $Spec(k)$-sheaf R^ip_*S is the Γ_k-module $H^i(\overline{X}, \overline{S})$ in the usual correspondence between sheaves of abelian groups over $Spec(k)$ and discrete Γ_k-modules ([Milne, EC], II, Thm. 1.9).

One can interpret the elements of $H^0(k, R^1p_*S)$ as Γ_k-invariant classes of $\overline{X}$-torsors under $\overline{S}$. They are given by torsors $\overline{Y} \to \overline{X}$ under $\overline{S}$ such that for any $g \in \Gamma_k$ the conjugate variety $^g\overline{Y}$ is isomorphic to $\overline{Y}$ as an $\overline{X}$-torsor under $\overline{S}$. (Recall that the conjugate variety $^g\overline{Y}$ is isomorphic to $\overline{Y}$ as a scheme, but the canonical morphism $^gq : {}^g\overline{Y} \to Spec(\overline{k})$ is $q : \overline{Y} \to Spec(\overline{k})$ followed by g^{*-1}.) This gives a collection of automorphisms ϕ_g of the scheme $\overline{Y}$ for all $g \in \Gamma_k$ such that $q \circ \phi_g = g^{*-1} \circ q$, and such that ϕ_g is compatible with the action of Γ_k on $\overline{X} = X \times_k \overline{k}$ via the second factor. Such a collection is a *descent datum* on the torsor $\overline{Y} \to \overline{X}$. It comes from an X-torsor $Y \to X$ under S if and only if the corresponding element of $H^0(k, R^1p_*S)$ belongs to the image of $H^1(X, S)$.

Proposition 2.2.4 *Suppose that a k-variety X and a commutative algebraic k-group S are such that $H^0(\overline{X}, \overline{S}) = \overline{S}(\overline{k})$, for instance, S is finite and X is geometrically connected, or S is of multiplicative type and $\overline{k}[X]^* = \overline{k}^*$, or S is affine and X is projective. If $X(k) \neq \emptyset$, then every descent datum on a torsor $\overline{Y} \to \overline{X}$ under $\overline{S}$ comes from a torsor $Y \to X$ under S.*

Proof. Under our assumption (2.14) can be written as

$$H^1(X, S) \longrightarrow H^0(k, R^1p_*S) \overset{\partial}{\longrightarrow} H^2(k, S) \overset{p^*}{\longrightarrow} H^2(X, S).$$

A k-point on X defines a section of the map p^*, which is therefore injective. Hence the map $H^1(X, S) \to H^0(k, R^1p_*S)$ is surjective. QED

We can regard this proposition as showing that the non-vanishing of $\partial(c)$ for some $c \in H^0(k, R^1p_*S)$ is an obstruction for the existence of a k-point on X.

2.3 Torsors under groups of multiplicative type

In the case of groups of multiplicative type the theory of torsors becomes transparent and more 'computable'. It is governed by the notion of 'type' of a torsor, introduced by Colliot-Thélène and Sansuc. The main simplifications are (assuming $\overline{k}[X]^* = \overline{k}^*$)

(1) over an algebraically closed field an $\overline{X}$-torsor is uniquely determined by its type,

(2) the set of isomorphism classes of X-torsors under S of any given type is either empty, or a principal homogeneous space under the group $H^1(k, S)$, and

(3) the unique obstruction for the existence of X-torsors of a given type lies in $H^2(k, S)$.

Type of a torsor.

We have seen in Lemma 2.2.2 that the group $\check{H}^1(Y/X, \mathbf{G}_m)$ can be identified with the group of equivalence classes of certain morphisms $Y \times_k S \to \mathbf{G}_m$, including all characters $S \to \mathbf{G}_m$. Let $c(\chi)$ be the class of a character $\chi \in \hat{S}(k)$ in $\check{H}^1(Y/X, \mathbf{G}_m)$.

Lemma 2.3.1 *Let X be a k-variety. Let $f : Y \to X$ be a torsor under a k-group S of multiplicative type, and let $\chi \in \hat{S}(k)$. The following elements of $Pic(X)$ coincide:*

(i) the class of the push-forward $\chi_*(Y) \to X$ in $H^1(X, \mathbf{G}_m)$,

(ii) the class of the subsheaf of χ-semiinvariants of the action of S on the Zariski sheaf $f_*(\mathcal{O}_Y)$ (up to sign),

(iii) the image of $c(\chi)$ under the map $\check{H}^1(Y/X, \mathbf{G}_m) \to H^1(X, \mathbf{G}_m)$ (cf. (2.3)).

Associating to $\chi \in \hat{S}(k)$ this class of $Pic(X)$ is a homomorphism of groups, functorial in k, X and S.

Proof. The equivalence of *(i)* and *(ii)* is immediate from the definition of push-forward and the relation between invertible sheaves and torsors under $\mathbf{G}_m$. The equivalence of *(i)* and *(iii)* follows from Lemma 2.2.2 for $G = S$ and $G' = \mathbf{G}_m$ by functoriality. Indeed, the following diagram commutes:

$$
\begin{array}{ccc}
\check{H}^1(Y/X, S) & \to & H^1(X, S) \\
\chi_* \downarrow & & \chi_* \downarrow \\
\check{H}^1(Y/X, \mathbf{G}_m) & \to & H^1(X, \mathbf{G}_m)
\end{array}
\tag{2.15}
$$

The last statement of Lemma 2.2.2 implies that χ_* sends the class of Y in $\check{H}^1(Y/X, S)$ to $c(\chi)$. This completes the proof. QED

Definition 2.3.2 *The homomorphism* $\hat{S} = \hat{S}(\overline{k}) \to Pic(\overline{X})$ *of Lemma 2.3.1 in the case of an algebraically closed field* $\overline{k}$ *is called* **the type** *of the torsor* $Y \to X$, *and is denoted by* type(Y).

When either $\overline{k}[Y]^* = \overline{k}^*$, or X, Y, and S are geometrically connected, type(Y) is the map $\hat{S}(\overline{k}) \to Pic(\overline{X})$ from (2.5) or (2.8), respectively. It follows from (i) that if $\overline{Y}$ is an $\overline{X}$-torsor under $\mathbf{G}_m$, then type($\overline{Y}$) sends the canonical generator $1 \in \mathbf{Z} = \hat{\mathbf{G}}_m$ to the class of this torsor in $Pic(\overline{X})$.

It is clear from the functoriality of type that if P is a k-torsor under S, then type($P \times_k X$) = 0.

Universal torsors.

Using the notion of type one defines another important notion which will be appearing in different places of this book. Recall that when X is a smooth and projective variety $Pic(\overline{X})$ is a finitely generated abelian group if and only if $H^1(\overline{X}, \mathcal{O}) = 0$.

Definition 2.3.3 (Colliot-Thélène–Sansuc) *Assume that* X *is a* k-*variety such that* $Pic(\overline{X})$ *is a finitely generated abelian group. A torsor* $Y \to X$ *under a group of multiplicative type* S *is called* **universal** *if its type* $h_{\overline{Y}}$ *is an isomorphism of* Γ_k-*modules* $\hat{S}$ *and* $Pic(\overline{X})$.

Note that there can be more than one type of universal torsors.

Suppose that $\overline{k}[X]^* = \overline{k}^*$. If a universal torsor $Y \to X$ exists, then *any* X-torsor under a k-group of multiplicative type S of *any* type $\lambda : \hat{S} \to Pic(\overline{X})$ is the push-forward $\lambda_*(Y)$, up to a twist by a k-torsor under S (this follows by functoriality from Corollary 2.3.9 below). Therefore, for such varieties X a universal torsor gives the finest possible partition of the set $X(k)$ among the torsors under groups of multiplicative type.

It is of course quite possible that X-torsors of some other type exist when the universal torsors do not. Another reason why we should not in general restrict ourselves to universal torsors is that they are not always as easy to describe by explicit equations as some other torsors (cf. Section 4.4). However, when explicit descriptions of universal torsors (or torsors closely related to them) are available, they are extremely useful. Universal torsors will appear in the appendix to Section 3.1 (X is a Del Pezzo surface of degree 5), at the end of Section 3.2 (X is a k-torsor under a semisimple k-group), in Section 6.3 (X is a compactification of a k-torsor under a torus), and in Section 9.5 (X is a homogeneous space of a semisimple group).

The spectral sequence of Ext's.

Let $p : X \to Spec(k)$ be the structure morphism.

Let M be a dicrete Γ_k-module identified with the corresponding sheaf on $Spec(k)$. The functor p_* from the category of étale sheaves on X to the category of sheaves over $Spec(k)$ has a left adjoint p^*, $Hom_k(M, p_*F) = Hom_{X_{\acute{e}t}}(p^*M, F)$. Since p^* is exact, we obtain the following Grothendieck spectral sequence (of composed functors p_* and $Hom_k(M, \cdot)$):

$$Ext_k^p(M, R^q p_* \mathbf{G}_m) \Rightarrow Ext_{X_{\acute{e}t}}^{p+q}(p^*M, \mathbf{G}_m). \qquad (2.16)$$

For $F = \mathbf{G}_m$ the exact sequence of low degree terms of (2.16) is

$$
\begin{aligned}
0 \longrightarrow \ & Ext_k^1(M, \overline{k}[X]^*) \to Ext_{X_{\acute{e}t}}^1(p^*M, \mathbf{G}_m) \to Hom_k(M, Pic(\overline{X})) \\
\overset{\partial}{\longrightarrow} \ & Ext_k^2(M, \overline{k}[X]^*) \to Ext_{X_{\acute{e}t}}^2(p^*M, \mathbf{G}_m).
\end{aligned}
$$
$$(2.17)$$

Let us define

$$e(X) := \partial(Id) \in Ext_k^2(Pic(\overline{X}), \overline{k}[X]^*)$$

for (2.17) with $M = Pic(\overline{X})$, and $Id \in Hom_k(Pic(\overline{X}), Pic(\overline{X}))$ the identity map. Let us note the following explicit interpretation of $e(X)$.

Theorem 2.3.4 (Colliot-Thélène–Sansuc) *(a) Let X be a smooth geometrically integral k-variety. The class $-e(X) \in Ext_k^2(Pic(\overline{X}), \overline{k}[X]^*)$ coincides with the class of the following natural 2-fold extension of Γ_k-modules:*

$$1 \to \overline{k}[X]^* \to \overline{k}(X)^* \to Div(\overline{X}) \to Pic(\overline{X}) \to 0. \qquad (2.18)$$

(b) Assume moreover that $\overline{k}[X]^* = \overline{k}^*$. Then the following properties are subject to the implications $(i) \Rightarrow (ii) \Leftrightarrow (iii) \Rightarrow (iv)$:

(i) $X(k) \neq \emptyset$,

(ii) $e(X) = 0$,

(iii) the natural map $\overline{k}^* \to \overline{k}(X)^*$ has a Γ_k-equivariant section,

(iv) the natural map $\overline{k}^* \to \overline{k}[U]^*$ has a section, where $U \subset X$ is a dense open set.

Proof. (a) The proof of the following fact can be found in ([Giraud], V.3.2.2) or, without using gerbs, and up to sign, in ([CS87a], Lemma 1.A.4). The statement seems to be well known, and the available proofs are somewhat involved, so we do not reproduce them here. Let

$$0 \to A \to B \to C \to 0$$

be an exact sequence of étale sheaves of abelian groups over $p : X \to Spec(k)$ such that the corresponding sequence

$$0 \to p_*(A) \to p_*(B) \to p_*(C) \to R^1 p_*(A) \to 0 \qquad (2.19)$$

is exact. Then the *transgression* map

$$Hom_k(\cdot, R^1 p_*(A)) \to Ext_k^2(\cdot, p_*(A))$$

from spectral sequence (2.16) is the inverse of the Yoneda pairing with the class of (2.19).

Let $X^{(1)}$ be the set of points of X of codimension 1,

$$Div_X = \oplus_{x \in X^{(1)}} i_{x*} \mathbf{Z}_x$$

be the sheaf of Weil divisors, where $i_{x*} : x \to X$ is the embedding of a point of codimension 1 into X. In our set-up we consider the exact sequence

$$0 \to \mathbf{G}_{m,X} \to j_* \mathbf{G}_{m,\eta} \to Div_X \to 0, \qquad (2.20)$$

where $j : \eta \to X$ is the embedding of the generic point. This sequence is exact since X is a smooth variety (hence Weil divisors coincide with Cartier divisors on X and its étale coverings; see [Milne, EC], II.3.9). Moreover, $(R^1 p_*)(j_* \mathbf{G}_{m,\eta})$ is a subsheaf of $R^1(pj)_* \mathbf{G}_{m,\eta}$ which is 0 by Hilbert's Theorem 90. Now applying p_* to (2.20) we get (2.18), and the statement *(a)* follows from the fact mentioned at the beginning of the proof.

(b) (i) $\Rightarrow$ (ii): The maps

$$Ext_k^i(M, \overline{k}^*) \to Ext_{X_{\acute{e}t}}^i(p^* M, \mathbf{G}_m), \quad i = 1, 2,$$

can be identified with p^*. A k-point of X defines a section of p^*, but by *(a)* we have $p^*(e(X)) = -p^*(\partial(Id)) = 0$ because (2.17) is a complex.

An independent proof of *(i) $\Rightarrow$ (iii)*: Take any point $P \in X(k)$. Let $\mathcal{O}_{\overline{X},P}$ be the local ring of $\overline{X}$ at P. Then the natural injection of Γ_k-modules $\overline{k}^* \hookrightarrow \mathcal{O}_{\overline{X},P}^*$ has a Γ_k-equivariant section s given by $g \mapsto g(P)$. On the other hand, the natural extension of Γ_k-modules

$$1 \to \mathcal{O}_{\overline{X},P}^* \to \overline{k}(X)^* \to Div_P(\overline{X}) \to 0,$$

where $Div_P(\overline{X}) \subset Div(\overline{X})$ consists of divisors passing through P, is split. Indeed, $Div_P(\overline{X})$ is a permutation Γ_k-module, hence

$$Ext_k^1(Div_P(\overline{X}), \mathcal{O}_{\overline{X},P}^*) = 0$$

by Shapiro's lemma and Hilbert's Theorem 90. Thus we have a Γ_k-equi-variant section $s' : \overline{k}(X)^* \to \mathcal{O}_{\overline{X},P}^*$. The composition ss' is a Γ_k-equivariant section of the map in *(iii)*.

The implication $(iii) \Rightarrow (ii)$ follows from *(a)*, and $(iii) \Rightarrow (iv)$ is obvious. $(ii) \Rightarrow (iii)$: The exact sequence (2.18) gives rise to the exact sequence

$$Ext_k^1(Div(\overline{X}), \overline{k}^*) \to Ext_k^1(\overline{k}(X)^*/\overline{k}^*, \overline{k}^*) \to Ext_k^2(Pic(\overline{X}), \overline{k}^*).$$

The first group is 0 by Shapiro's lemma and Hilbert's Theorem 90. Thus the second arrow in the above sequence is injective. But this arrow is precisely the Yoneda pairing with the class of the extension

$$1 \to \overline{k}(X)^*/\overline{k}^* \to Div(\overline{X}) \to Pic(\overline{X}) \to 1,$$

and it sends the class of the extension

$$1 \to \overline{k}^* \to \overline{k}(X)^* \to \overline{k}(X)^*/\overline{k}^* \to 1$$

to $e(X)$. This proves the implication. QED

Definition 2.3.5 (Colliot-Thélène–Sansuc) *Let X be a k-variety such that $\overline{k}[X]^* = \overline{X}^*$. The class $e(X)$, which by Theorem 2.3.4 (b) is an obstruction to the existence of a k-point on X, is called* **the elementary obstruction**.

The fundamental exact sequence.

From now on let M be a Γ_k-module finitely generated as an abelian group, and let $S = Hom_{k\text{-groups}}(M, \mathbf{G}_m)$ be the k-group of multiplicative type dual to M (in the category of commutative algebraic k-groups). Then $M = Hom_{k\text{-groups}}(S, \mathbf{G}_m)$ is the character group of S, and is denoted by $M = \hat{S}$. If M is finite, then S is also finite, and if M is torsion free, then S is an algebraic k-torus.

If the torsion of M is coprime with the characteristic of k, then S is smooth over k, and in what follows the flat topology can be replaced by the étale topology. Without this additional assumption the groups $H^i(X, S)$ in the following theorem must be taken with respect to the flat topology.

Theorem 2.3.6 (Colliot-Thélène–Sansuc) *There is the following exact sequence ('the fundamental exact sequence'):*

$$0 \longrightarrow Ext_k^1(M, \overline{k}[X]^*) \to H^1(X, S) \xrightarrow{\text{type}} Hom_k(M, Pic(\overline{X}))$$
$$\xrightarrow{\partial} Ext_k^2(M, \overline{k}[X]^*) \to H^2(X, S).$$

$$(2.21)$$

The map type *associates to a torsor its type.*

Proof. This is a particular case of (2.17) in view of the following general fact.

Lemma 2.3.7 *Let S be an X-group scheme of multiplicative type. Then we have an isomorphism*

$$H^i(X, S) = Ext^i_{X_{\acute{e}t}}(\hat{S}, \mathbf{G}_m)$$

functorial in S and X.

Here $H^i(X, S)$ is taken with respect to the *fppf*-topology. Again, when S is smooth, *fppf* can be replaced by *ét*.

To prove the last assertion of the theorem we may assume that $k = \overline{k}$ and $S = \mathbf{G}_m$, then the map labelled `type` in (2.21) is clearly the identity map $Pic(\overline{X}) \rightarrow Pic(\overline{X})$, and hence gives the type of the torsor. QED

Sublemma 2.3.8 *We have $\mathcal{E}xt^i_{X_{fppf}}(p^*M, \mathbf{G}_m)) = 0$ for $i > 0$. If the torsion of M is coprime with the characteristic of k, then X_{fppf} can be replaced by $X_{\acute{e}t}$.*

Proof. Locally, M is a finite direct sum of copies of $\mathbf{Z}$ and $\mathbf{Z}/n$. Let X be the spectrum of a strictly henselian local ring, then we have

$$Ext^i_{X_{fppf}}(\mathbf{Z}, \mathbf{G}_m) = Ext^i_{X_{fppf}}(\mathbf{Z}/n, \mathbf{G}_m) = 0$$

for any $i > 0$. The vanishing of the first group is obvious (here $\mathbf{G}_m$ can be replaced by any sheaf). The vanishing of the second group for $i > 1$ follows from the exact sequence $0 \rightarrow \mathbf{Z} \rightarrow \mathbf{Z} \rightarrow \mathbf{Z}/n \rightarrow 0$. Then multiplication by n defines a surjection of sheaves for the flat topology $\mathbf{G}_m \rightarrow \mathbf{G}_m$. The fact that this map is surjective is equivalent to the vanishing of $Ext^1_{X_{fppf}}(\mathbf{Z}/n, \mathbf{G}_m)$. QED

Proof of Lemma 2.3.7. The local-to-global spectral sequence ([Grothendieck, SGA 4], V.6.1)

$$H^p(X_{fppf}, \mathcal{E}xt^q_{X_{fppf}}(M, \mathbf{G}_m)) \Rightarrow Ext^{p+q}_{X_{fppf}}(\hat{S}, \mathbf{G}_m)$$

completely degenerates by the previous sublemma. We can take the edge map of this spectral sequence as the isomorphism of the lemma, provided we show that $Ext^i_{X_{\acute{e}t}}(\hat{S}, \mathbf{G}_m) = Ext^i_{X_{fppf}}(\hat{S}, \mathbf{G}_m)$. Since $\hat{S}$ is locally constant for the étale topology, by the Leray spectral sequence of a covering ([Grothendieck, SGA 4], V.3.3) it is enough to check the cases $\hat{S} = \mathbf{Z}$ and $\hat{S} = \mathbf{Z}/n$. In the first case, the formula to be justified is $H^i(X_{\acute{e}t}, \mathbf{G}_m) = H^i(X_{fppf}, \mathbf{G}_m)$ which is true since $\mathbf{G}_m$ is smooth. The second case follows from this by the lemma of five homomorphisms. QED

Let $\lambda : M \to Pic(\overline{X})$ be a Γ_k-homomorphism. The exact sequence (2.21) implies that an X-torsor under S of type λ exists if and only if $\partial(\lambda) = 0$. If $Pic(\overline{X})$ is of finite type, then, by definition, universal X-torsors are torsors under S whose type is an isomorphism $\hat{S} \simeq Pic(\overline{X})$. It follows from (2.21) that *universal X-torsors* (of any 'universal' type) *exist if and only if the elementary obstruction vanishes on X*, that is, $\partial(Id) = e(X) = 0$.

Corollary 2.3.9 *Let X satisfy the property that the only invertible functions on $\overline{X}$ are constants, that is, $\overline{k}[X]^* = \overline{k}^*$ (in particular, X is geometrically connected). Then we have the exact sequence*

$$0 \longrightarrow H^1(k,S) \xrightarrow{p^*} H^1(X,S) \xrightarrow{\text{type}} Hom_k(M, Pic(\overline{X}))$$
$$\xrightarrow{\partial} H^2(k,S) \xrightarrow{p^*} H^2(X,S).$$
$$(2.22)$$

For $M = \mathbf{Z}$ the spectral sequence (2.17) gives rise to the exact sequence

$$0 \to Pic(X) \to Pic(\overline{X})^{\Gamma_k} \xrightarrow{\partial} Br(k) \xrightarrow{p^*} Br_1(X)$$
$$\xrightarrow{r} H^1(k, Pic(\overline{X})) \to H^3(k, \mathbf{G}_m) \xrightarrow{p^*} H^3(X, \mathbf{G}_m)$$
$$(2.23)$$

Proof. We apply Lemma 2.3.7 to $Spec(k)$. To obtain (2.23) one uses Hilbert's Theorem 90. The identification of the arrows marked p^* follows from the functoriality of the isomorphism of Lemma 2.3.7 which implies the commutativity of the following diagram:

$$\begin{array}{ccc} Ext_k^2(M, \overline{k}^*) & \to & H^2(k,S) \\ \| & & p^* \downarrow \\ Ext_k^2(M, \overline{k}^*) & \to & H^2(X,S) \end{array}$$
$$(2.24)$$

Alternatively, this follows from Proposition 2.3.11 below, taking into account that $\overline{k}[X]^* = \overline{k}^*$ implies $S = p_* p^* S$. QED

Note that the assumption of the corollary is satisfied by all proper geometrically connected and geometrically reduced varieties.

It is clear from (2.22) that for any homomorphism of abelian groups $\lambda : M \to Pic(\overline{X})$ there exists exactly one $\overline{X}$-torsor under $\overline{S}$ of type λ. Thus the type of a torsor $Y \to X$ under S and the torsor $\overline{Y} \to \overline{X}$ under $\overline{S}$ uniquely determine each other. This is property (1) formulated at the beginning of this section. The other two are obvious from (2.22).

For varieties X satisfying $\overline{k}[X]^* = \overline{k}^*$ one observes that if $X(k) \neq \emptyset$, then the maps p^* in (2.22) and (2.23) have a section given by the specialization at a k-point. In this case $\partial(\lambda) = 0$ for any λ. Though obvious this is an important observation: it shows that *the existence of X-torsors under groups of multiplicative type is a necessary condition for the existence of a k-point on X.*

It follows by functoriality from the proposition above that

$$\partial(\lambda) = \lambda^*(e(X)).$$

The vanishing of this class is a necessary and sufficient condition for the existence of X-torsors of type λ. The following corollary gives a necessary and sufficient condition for the existence of torsors of a frequently used type.

Corollary 2.3.10 *Let X be a smooth and geometrically integral k-variety such that $\overline{k}[X]^* = \overline{k}^*$. Let $U \subset X$ be a dense open set, and let $\lambda \in Hom_k(\hat{S}, Pic(\overline{X}))$ be an injective map whose image is generated by the classes of divisors supported in $\overline{X} \setminus \overline{U}$. Then X-torsors of type λ exist if and only if the exact sequence of Γ_k-modules*

$$1 \to \overline{k}^* \to \overline{k}[U]^* \to \overline{k}[U]^*/\overline{k}^* \to 1 \qquad (2.25)$$

is split.

Proof. It follows from Theorem 2.3.4 *(a)* and from the functoriality of (2.22) that the inverse of the class $\partial(\lambda) = \lambda^*(e(X))$ is represented by the 2-fold extension

$$1 \to \overline{k}^* \to \overline{k}[U]^* \to Div_{\overline{X} \setminus \overline{U}}(\overline{X}) \to \hat{S} \to 0.$$

To show that this extension splits if and only if (2.25) does, we apply the same argument as in the proof of the implication $(ii) \Rightarrow (iii)$ of Theorem 2.3.4 *(b)*. QED

By (2.22) a torsor of a given type, if it exists, is unique up to twist by a cocycle of $H^1(k, S)$. Therefore we have

$$X(k) = \bigcup_{\text{type}(Y,f)=\lambda} f(Y(k)). \qquad (2.26)$$

This formula also makes sense when either $X(k)$ or the set of torsors of type λ is empty.

Comparison of two spectral sequences.

The following proposition compares the spectral sequence of Ext's with the Leray spectral sequence.

Proposition 2.3.11 *Let S be a k-group of multiplicative type. There is a canonical functorial morphism of the Leray spectral sequence (2.13) to the spectral sequence (2.16):*

$$E_2^{p,q} = H^p(k, H^q(\overline{X}, \overline{S})) \longrightarrow E'^{p,q}_2 = Ext_k^p(\hat{S}, R^q p_* \mathbf{G}_m),$$

both sequences converging to $H^i(X,S)$. When S is a torus, this morphism is an isomorphism. If S is an arbitrary k-group of multiplicative type, and X is a k-variety such that $\overline{k}[X]^*$ is divisible (for example, $\overline{k}[X]^* = \overline{k}^*$), then this morphism induces an isomorphism of the exact sequences of the first five low degree terms. In particular, we have a canonical isomorphism $H^0(k, H^1(\overline{X}, \overline{S})) = Hom_k(\hat{S}, Pic(\overline{X}))$ associating to a torsor its type.

Proof. Let $M = \hat{S}$. The sheaf of abelian groups on X given by S can be written as $\mathcal{H}om_X(p^*M, \mathbf{G}_m)$ (M is a Γ_k-module of finite type, and checking it locally one sees that here homomorphisms of sheaves over X are the same thing as morphisms of group X-schemes). We have

$$p_* \mathcal{H}om_X(p^*M, \mathcal{F}) = \mathcal{H}om_k(M, p_* \mathcal{F})$$

since p^* and p_* are adjoint (this is a sheaf version of $Hom_X(p^*M, \mathcal{F}) = Hom_k(M, p_* \mathcal{F})$). This implies that there is a natural isomorphism of functors from the derived category $\mathcal{D}^+(X)$ of sheaves on $X_{\acute{e}t}$ to the derived category $\mathcal{D}^+(k)$ of sheaves on $Spec(k)$ (complexes of Γ_k-modules)

$$\mathbf{R}p_* \mathbf{R}\mathcal{H}om_X(p^*M, \cdot) = \mathbf{R}\mathcal{H}om_k(M, \mathbf{R}p_*(\cdot)). \tag{2.27}$$

Let us apply this to the sheaf $\mathbf{G}_m$. We have $\mathcal{E}xt_X^i(p^*M, \mathbf{G}_m) = 0$ for $i > 0$, by Sublemma 2.3.8, therefore the complex $\mathbf{R}\mathcal{H}om_X(p^*M, \mathbf{G}_m)$ consists of the sheaf

$$S = \mathcal{H}om_X(p^*M, \mathbf{G}_m)$$

in degree 0. We obtain

$$\mathbf{R}p_*(S) = \mathbf{R}\mathcal{H}om_k(M, \mathbf{R}p_* \mathbf{G}_m). \tag{2.28}$$

Now we apply the derived functor $\mathbf{H}(k, \cdot)$ of the functor $A \mapsto A^{\Gamma_k}$ to both sides of (2.28). On the left hand side we get $\mathbf{H}(k, \mathbf{R}p_*S) = \mathbf{H}(X, S)$ (where $\mathbf{H}(X, \cdot)$ is the derived functor of $H^0(X, \cdot)$). The resulting spectral sequence of composed functors is

$$H^p(k, R^q p_* S) \Rightarrow H^{p+q}(X, S).$$

On the right hand side we get $\mathbf{R}\mathcal{H}om_k(M, \mathbf{R}p_* \mathbf{G}_m)$ which gives rise to the spectral sequence

$$Ext_k^p(M, R^q p_* \mathbf{G}_m) \Rightarrow H^{p+q}(X, S).$$

Let us relate these spectral sequences.

Let $\mathcal{G}$ be a complex of sheaves on $X_{\acute{e}t}$ which is an injective resolution of $\mathbf{G}_m$. There is a natural map of complexes of Γ_k-modules (=sheaves over $Spec(k)$)

$$p_*\mathcal{H}om_X(p^*M, \mathcal{G}) \to \mathcal{H}om_k(M, p_*\mathcal{G}). \qquad (2.29)$$

Let $I = I^{**}$ be a bicomplex which is a Cartan–Eilenberg resolution of $p_*\mathcal{G}$, and let $\mathcal{T}$ be the total complex of $\mathcal{H}om_k(M, I)$. The complex $\mathcal{T}$ represents $\mathbf{R}\mathcal{H}om_k(M, p_*\mathbf{G}_m)$ in $\mathcal{D}^+(k)$ ([Weibel], 10.5.6). The resolution $p_*\mathcal{G} \to I$ induces a natural map of complexes $\mathcal{H}om_k(M, p_*\mathcal{G}) \to \mathcal{T}$, and on combining with (2.29) a map

$$p_*\mathcal{H}om_X(p^*M, \mathcal{G}) \to \mathcal{T}. \qquad (2.30)$$

This gives a map between the corresponding hypercohomology spectral sequences, which is the natural morphism we are looking for. (We have actually proved more: the image of this morphism is contained in the image of the natural map $H^p(k, \mathcal{H}om_k(M, R^q p_*\mathbf{G}_m)) \to Ext_k^p(M, R^q p_*\mathbf{G}_m)$.)

If S is a torus, then M is locally in the étale topology isomorphic to a finite direct sum of copies of $\mathbf{Z}$. Then (2.29) is an identity map, and the remaining assertions of the proposition become obvious.

If S is any k-group of multiplicative type it is enough to prove that (2.30) induces isomorphisms on H^0 and H^1 (in other words, (2.30) induces a quasi-isomorphism of complexes truncated at 1). We can check this over $\bar{k}$, and hence suppose that $M = \mathbf{Z}$ or $M = \mathbf{Z}/n$. The first case is already clear. Let $M = \mathbf{Z}/n$, then for H^0 we have $H^0(\overline{X}, \mu_n) \to Hom(\mathbf{Z}/n, H^0(\overline{X}, \mathbf{G}_m))$ which is always an isomorphism. For H^1 we get the natural map

$$H^1(\overline{X}, \mu_n) = H^1(\overline{X}, Hom(\mathbf{Z}/n, \mathbf{G}_m)) \to Hom(\mathbf{Z}/n, H^1(\overline{X}, \mathbf{G}_m)).$$

From the definition of type (cf. Lemma 2.3.1 (i)) and functoriality it is clear that this map sends a torsor to its type. This map is always surjective, and its kernel is $H^0(\overline{X}, \mathbf{G}_m)/n$ which is trivial under our assumptions. This finishes the proof of the proposition. QED

The advantage of the spectral sequence of Ext's (2.16) is that it is more handy in applications (obtaining torsors from universal ones, restricting to an open subset, etc.). The Leray spectral sequence is more general; in particular, the transgression map $H^0(k, R^1 p_*S) \to H^2(k, p_*S)$ can produce a non-trivial cohomology class even when $Pic(\overline{X}) = 0$ and the corresponding map of the sequence of Ext's is zero. Using the Leray spectral sequence one can interpret the obstruction for the existence of a torsor of a given

type as the obstruction for a 'descent datum' on an $\overline{X}$-torsor to come from
an X-torsor (cf. Proposition 2.2.4).

2.4 Obstructions to existence of rational points over arbitrary fields

In this section we review some other known obstructions for the existence
of rational points on varieties over an arbitrary field k. They appear as
obstructions for the splitting of certain natural extensions of the Galois
group Γ_k. The most important one is given by the exact sequence of the
étale fundamental group of X constructed by Grothendieck in [Grothen-
dieck, SGA 1]. We relate it to the elementary obstruction $e(X)$ (Definition
2.3.5).

Let k be a field of characteristic 0. In this section we denote by $\overline{Y}, \overline{Z}, \overline{U}$,
etc. the varieties over $\overline{k}$, and this notation should not be understood in the
sense that these necessarily come from k-varieties. This possibly could be
the case, so if X is a variety over k, we have our usual notation $\overline{X} = X \times_k \overline{k}$.
We write $X_K = X \times_k K$ for an intermediate extension $k \subset K \subset \overline{k}$.

For a morphism of schemes $Y \to X$ let $Aut(Y/X)$ be the group of X-
automorphisms of Y. If $\overline{p} : \overline{Y} \to Spec(\overline{k})$ is a connected and reduced
$\overline{k}$-variety the integral closure of k in $\overline{k}[\overline{Y}]$ is $\overline{k}$. Then any k-automorphism
of $\overline{Y}$ induces an automorphism g of $\overline{k}$ over k. Such a g is then unique, thus
we have a natural group homomorphism $Aut(\overline{Y}/k) \to \Gamma_k$.

(1) Let X be a geometrically integral k-variety, $\overline{x} \in X(\overline{k})$. In [Gro-
thendieck, SGA 1] Grothendieck proved that there is an exact sequence of
profinite groups

$$1 \to \pi_1(\overline{X}, \overline{x}) \to \pi_1(X, \overline{x}) \to \Gamma_k \to 1 \qquad\qquad (\pi_X)$$

which is split if $X(k) \neq \emptyset$ by the functoriality of the fundamental group
(one then chooses $\overline{x}$ defined by a k-point). We shall refer to the equivalence
class of the extension (π_X) as the *fundamental* obstruction.

(2) Let X be a homogeneous space of a connected algebraic group H.
Choose $\overline{x} \in X(\overline{k})$, and let $\overline{G} \subset \overline{H}$ be the stabilizer of $\overline{x}$. Springer in
[Springer 66] introduced the obstruction for lifting X to a k-torsor under
H. He presented this obstruction as a class in the second Galois cohomology
group with coefficients in an appropriate kernel (=lien) on $\overline{G}(\overline{k})$ ('kernel of
stabilizers'). Equivalently, Springer's cohomology class is the class of the

extension

$$1 \to \overline{G}(\overline{k}) \to E \to \Gamma_k \to 1 \qquad (2.31)$$

where E can be chosen to be the subgroup of the semidirect product of $H(\overline{k})$ and Γ_k consisting of products hg, $h \in H(\overline{k})$, $g \in \Gamma_k$, such that ${}^g\overline{x} = \overline{x}h$ (see [FSS], (5.1), and Section 9.2 in the last chapter of this book).

In fact, Springer shows that there always exist a homogeneous space X' of H and a surjective morphism $X' \to X$ compatible with the action of H, such that the stabilizers of $\overline{k}$-points of X' are finite and nilpotent ([Springer 66], Thm. 3.8). Therefore Springer's obstruction can be considered as a particular case of the fundamental obstruction. More precisely, (2.31) with $\overline{G}$ finite and nilpotent ia a push-out of (π_X).

Let S be a finite k-group, and $\overline{Y} \to \overline{X}$ be a right torsor under $\overline{S}$ such that $\overline{Y}$ is integral. Choosing a point $\overline{y} \in \overline{Y}(\overline{k})$ above $\overline{x}$ we realize $Aut(\overline{Y}/X)$ (resp. $S(\overline{k}) = Aut(\overline{Y}/\overline{X})$) as a quotient of $\pi_1(X, \overline{x})$ (resp. of $\pi_1(\overline{X}, \overline{x})$). We have an exact sequence

$$1 \to S(\overline{k}) = Aut(\overline{Y}/\overline{X}) \to Aut(\overline{Y}/X) \to \Gamma_k. \qquad (2.32)$$

This sequence is right exact if and only if the kernel of the corresponding map $\pi_1(\overline{X}, \overline{x}) \to S(\overline{k})$, which is normal in $\pi_1(\overline{X}, \overline{x})$, is also stable under the action of Γ_k by outer automorphisms. In other words, this kernel should also be normal in $\pi_1(X, \overline{x})$. This is equivalent to the condition that all the conjugate varieties of $\overline{Y}$ are isomorphic as $\overline{X}$-torsors under $\overline{S}$. When (2.32) is exact, the splitting of this sequence is a weaker but more explicit obstruction to the existence of k-points on X than the fundamental obstruction.

We now relate the abelianized fundamental obstruction to the obstructions discussed in Section 2.3.

Let U be a geometrically integral variety over k. Let $\pi^{ab}(\overline{U})$ be the abelianization of $\pi_1(\overline{U}, \overline{u})$ in the category of profinite groups. (We omit the base point since the abelianized fundamental groups for different choices of the base point are canonically isomorphic.) Consider the push-out of (π_U) with respect to the abelianization map $\pi_1(\overline{U}, \overline{u}) \to \pi_1^{ab}(\overline{U})$:

$$1 \to \pi_1^{ab}(\overline{U}) \to P \to \Gamma_k \to 1. \qquad (\pi_U^{ab})$$

The abelianized fundamental group $\pi_1^{ab}(\overline{U})$ with its Γ_k-module structure can be computed as follows:

$$\pi_1^{ab}(\overline{U}) = \varprojlim \pi_1^{ab}(\overline{U})/n,$$

where $\pi_1^{ab}(\overline{U})/n = Hom(H^1(\overline{U},\mu_n),\overline{k}^*)$ (see, e.g., [KL], (1.1)). The Kummer sequence gives an exact sequence of Γ_k-modules

$$0 \to \overline{k}[U]^*/\overline{k}[U]^{*n} \to H^1(\overline{U},\mu_n) \to Pic(\overline{U})[n] \to 0.$$

The map $\overline{k}^* \to \overline{k}^*$ given by $x \mapsto x^n$ is surjective, hence we have

$$\overline{k}[U]^*/\overline{k}[U]^{*n} = (\overline{k}[U]^*/\overline{k}^*)/n.$$

We obtain a natural map of Γ_k-modules

$$\pi_1^{ab}(\overline{U})/n = Hom(H^1(\overline{U},\mu_n),\overline{k}^*) \to Hom((\overline{k}[U]^*/\overline{k}^*)/n,\overline{k}^*)$$

It is surjective because $Ext_{\mathbf{Z}}^1(\cdot,\overline{k}^*) = 0$ which is due to the well known fact that $\overline{k}^*$ is divisible. Therefore we get a natural surjective map of Γ_k-modules

$$\alpha_n : \pi_1^{ab}(\overline{U}) \to Hom((\overline{k}[U]^*/\overline{k}^*)/n,\overline{k}^*). \tag{2.33}$$

Theorem 2.4.1 *Let U be a geometrically integral k-variety. The following conditions are equivalent:*
(i) the exact sequence of Γ_k-modules (2.25) is split,
(ii) the pushed-out extension $\alpha_{n*}(\pi_U^{ab})$ is split for all n.

Corollary 2.4.2 *Let X be a smooth and geometrically integral variety over k such that $\overline{k}[X]^* = \overline{k}^*$, and $Pic(\overline{X})$ is a finitely generated abelian group. Let S be the k-group of multiplicative type dual to $Pic(\overline{X})$. Let $U \subset X$ be a dense open subset such that $Pic(\overline{U}) = 0$. Then $e(X) = 0$ if and only if the push-out of the extension (π_U^{ab}) by the maps $\pi^{ab}(\overline{U}) \to \pi^{ab}(\overline{U})/n$ is split for all n.*

To derive Corollary 2.4.2 from Theorem 2.4.1 observe that since $Pic(\overline{U}) = 0$, the torsors considered in Corollary 2.3.10 are universal. Thus Corollary 2.3.10 says that (2.25) splits if and only if universal X-torsors exist, and that is equivalent to the vanishing of the elementary obstruction $e(X) = 0$. We also note that $Pic(\overline{U}) = 0$ implies that $\pi^{ab}(\overline{U})/n = Hom((\overline{k}[U]^*/\overline{k}^*)/n,\overline{k}^*)$.

Proof of Theorem 2.4.1. Let R be the k-torus dual to the torsion free Γ_k-module $\overline{k}[U]^*/\overline{k}^*$. We rewrite (2.25) as an extension of the Γ_k-modules $\hat{R}$ by $\overline{k}^*$:

$$1 \to \overline{k}^* \to \overline{k}[U]^* \to \hat{R} \to 1. \tag{$*_U$}$$

It is clear that a k-point of U (actually, even a 0-cycle of degree 1 on U) defines a splitting of $(*_U)$. Hence the non-vanishing of the class of this extension $[*_U] \in Ext^1_k(\hat{R}, \overline{k}^*) = H^1(k, R)$ is an obstruction to the existence of k-points on U.

The multiplication by n sequence $1 \to R[n] \to R \to R \to 1$ defines the boundary maps $\partial_n : H^1(k, R) \to H^2(k, R[n])$. We thus get a family of elements $\partial_n([*_U]) \in H^2(k, R[n])$.

In our notation α_n is the map from $\pi_1^{ab}(\overline{U})$ to $Hom(\hat{R}/n, \overline{k}^*) = R(\overline{k})[n]$. We define $\overline{Z}_n \to \overline{U}$ as the connected étale covering corresponding to the group homomorphism $\pi_1(\overline{U}, \overline{u}) \to R(\overline{k})[n] = Aut(\overline{Z}_n/\overline{U})$ with Galois invariant kernel. We get a commutative diagram of group extensions

$$
\begin{array}{ccccccccc}
1 & \to & R(\overline{k})[n] & \to & Aut(\overline{Z}_n/U) & \to & \Gamma_k & \to & 1 \\
& & \uparrow & & \uparrow & & \| & & \\
1 & \to & \pi_1(\overline{U}, \overline{u}) & \to & \pi_1(U, \overline{u}) & \to & \Gamma_k & \to & 1
\end{array}
\qquad (2.34)
$$

Note that the Γ_k-module structure on $R(\overline{k})[n]$ given by the upper row of (2.34) is its usual Γ_k-module structure; the class of this extension is $\alpha_{n*}(\pi_U^{ab})$.

Claim. *The class in $H^2(k, R[n])$ of the upper extension (2.34) coincides with $\partial_n([*_U])$.*

Theorem 2.4.1 follows from this claim since $H^1(k, R)$ contains no divisible elements. (Indeed, let K be a finite extension of k over which R is isomorphic to $\mathbf{G}^r_m$. Now $H^1(K, \mathbf{G}^r_m) = 0$ by Hilbert's Theorem 90, and we conclude by a restriction–corestriction argument.)

Before proving our claim we state a few useful lemmas (borrowed from [Manin, CF], IV.8).

Lemma 2.4.3 *Let Y be a k-torsor under a k-torus R. Let $[Y] \in H^1(k, R)$ be the class of Y. There is a canonical isomorphism of Γ_k-modules $\overline{k}[Y]^*/\overline{k}^*$ and $\hat{R}$. The class of the extension*

$$
1 \to \overline{k}^* \to \overline{k}[Y]^* \to \hat{R} \to 0 \qquad (*_Y)
$$

in $Ext^1_k(\hat{R}, \overline{k}^*) = H^1(k, Hom(\hat{R}, \overline{k}^*)) = H^1(k, R)$ is the inverse of $[Y]$.

Proof. By Rosenlicht's lemma any invertible function on $\overline{R}$ is a character multiplied by a constant, so that $\overline{k}[R]^*/\overline{k}^* = \hat{R}$. Let Y be given by a cocycle $\sigma \in Z^1(\Gamma_k, R(\overline{k}))$ so that the (twisted) action of $g \in \Gamma_k$ on $s \in Y(\overline{k}) = R(\overline{k})$ is $g(s) = \sigma(g) \cdot {}^g s$. The translations by elements of $R(\overline{k})$ act trivially on $\overline{k}[R]^*/\overline{k}^* = \hat{R}$ (the translation by s multiplies the character by its value at s). Hence an isomorphism $\overline{k}[Y]^*/\overline{k}^* = \overline{k}[R]^*/\overline{k}^*$ does not depend on the trivialization of Y and is Γ_k-equivariant.

More precisely, the twisted action of Γ_k induces the following action on a character $\chi \in \hat{R}$ considered as an invertible function on $\overline{Y}$: $g(\chi) = {}^g\chi \cdot {}^g\chi(\sigma(g)^{-1})$. On the other hand, the cocycle corresponding to $(*_Y)$ can be given by $\chi(s)/{}^g\chi(g^{-1}(s)) = \chi(s)/{}^g\chi(\sigma(g)^{-1} \cdot {}^{g^{-1}}s)$. Evaluating this constant on the neutral element of R we see that the last expression is just ${}^g\chi(\sigma(g))$. Thus the cohomology class of extension $(*_Y)$ is the inverse of the class of Y. QED

Lemma 2.4.4 *Let Y be a k-torsor under R such that $[Y] = -[*_U] \in H^1(k, R)$. Then there exists a morphism $q : U \to Y$ such that $q^* : \overline{k}[Y]^* \to \overline{k}[U]^*$ is an isomorphism. In particular, the morphism of $(*_Y)$ onto $(*_U)$ defined by q^* is an equivalence of group extensions of Γ_k by $\overline{k}^*$. Any morphism of U to a k-torsor under a torus factors through q. If $\overline{U} \simeq \mathbf{G}_m^n$, then q is an isomorphism, hence U is isomorphic to a k-torsor under a torus.*

Proof. By Lemma 2.4.3 we have $[*_U] = [*_Y]$. Therefore there exists an isomorphism of Γ_k-modules $\rho : \overline{k}[Y]^* \to \overline{k}[U]^*$ such that we have a commutative diagram

$$
\begin{array}{ccccccccc}
1 & \to & \overline{k}^* & \to & \overline{k}[U]^* & \to & \hat{R} & \to & 1 \\
  &   & \| &   & \uparrow &   & \| &   & \\
1 & \to & \overline{k}^* & \to & \overline{k}[Y]^* & \to & \hat{R} & \to & 1
\end{array}
$$

It is a general fact that $\rho \in Hom_k(\overline{k}[Y]^*, \overline{k}[U]^*)$ (homomorphisms of Γ_k-modules) uniquely defines $\tilde{\rho} \in Hom_{\Gamma_k, \overline{k}\text{-alg}}(\overline{k}[Y], \overline{k}[U])$ (Γ_k-equivariant homomorphisms of $\overline{k}$-algebras) such that $\tilde{\rho}$ gives ρ on invertible elements. (Indeed, as a $\overline{k}$-variety $\overline{Y}$ is isomorphic to a torus $\mathbf{G}_m^r$. Therefore, $\overline{k}[Y]^*$ generates the $\overline{k}$-algebra $\overline{k}[Y]$, hence $\tilde{\rho}$ is uniquely determined by ρ. To show that for any ρ there exists some $\tilde{\rho}$ we reason as follows. A $\overline{k}$-point of Y realizes $\hat{R}$ inside $\overline{k}[Y]^*$ as functions χ which equal 1 at this point. As a $\overline{k}$-vector space, $\overline{k}[Y]$ is freely generated by the characters $\chi \in \hat{R}$. Thus ρ restricted to the subgroup $\hat{R} \subset \overline{k}[Y]^*$ is a homomorphism $\hat{R} \to \overline{k}[U]^*$. It uniquely extends to a morphism of $\overline{k}$-algebras $\tilde{\rho} : \overline{k}[Y] \to \overline{k}[U]$; then $\tilde{\rho}$ restricted to $\overline{k}[Y]^*$ is just ρ. It is clear that $\tilde{\rho}$ is Γ_k-equivariant since so is its restriction to $\overline{k}[Y]^*$.) Now we define $q : U \to Y = Spec(k[Y])$ as the morphism dual to the morphism of k-algebras $k[Y] = \overline{k}[Y]^{\Gamma_k} \to \overline{k}[U]^{\Gamma_k} = k[U]$ defined by $\tilde{\rho}$.

If $q' : U \to Y'$ is another morphism to a k-torsor under a k-torus R', then q'^* sends $(*_{Y'})$ to $(*_U)$. Then $(q^*)^{-1}q'^*$ sends $(*_{Y'})$ to $(*_Y)$. Dualizing the corresponding map of Γ_k-modules $\hat{R}' \to \hat{R}$ we get a morphism of tori $\xi : R \to R'$. The push-forward torsor, given by the contracted product

$Y \times^R R' = (Y \times_k R')/R$, is canonically isomorphic to Y'. Thus we have a morphism of torsors $\tau : Y \to Y'$ compatible with the action of R via ξ. To prove that $q' = \tau q$ it is enough to prove that $q'^* = q^* \tau^*$ for the $\overline{k}$-algebras of regular functions. This fact is clear since it holds for the corresponding groups of invertible functions.

Now assume that $\overline{U} \simeq \mathbf{G}_m^n$. To show that q is an isomorphism, it is enough to prove this over $\overline{k}$. A morphism of affine varieties $\overline{U} \to \overline{Y}$ is an isomorphism if and only if the dual map $\tilde{\rho} : \overline{k}[Y] \to \overline{k}[U]$ is an isomorphism of $\overline{k}$-algebras. After choosing base points both algebras considered as $\overline{k}$-vector spaces are freely generated by the characters of R. By construction, $\tilde{\rho}$ induces an isomorphism on $\hat{R}$, hence $\tilde{\rho}$ is an isomorphism. QED

Lemma 2.4.5 *Let* $1 \to A \to B \to C \to 1$ *be a central extension of algebraic k-groups such that B and C are geometrically connected and A finite. Then $A(\overline{k}) = \mathrm{Aut}(\overline{B}/\overline{C})$. Let Y be a right k-torsor under C. Choose a base point $\overline{y}_0 \in Y(\overline{k})$, and let $\nu : \overline{C} \to \overline{Y}$ be the isomorphism of right torsors under $\overline{C}$ sending the neutral element to $\overline{y}_0$. Let $\overline{B} \to \overline{Y}$ be the composition of $\overline{B} \to \overline{C}$ with ν. Then we have the extension of groups*

$$1 \to A(\overline{k}) = \mathrm{Aut}(\overline{B}/\overline{Y}) \to \mathrm{Aut}(\overline{B}/Y) \to \Gamma_k \to 1, \qquad (2.35)$$

such that the induced Γ_k-module structure on $A(\overline{k})$ is its usual Γ_k-module structure. The class of (2.35) in $H^2(k, A)$ coincides with $\partial([Y])$, where ∂ is the connecting homomorphism $H^1(k, C) \to H^2(k, A)$.

Proof. A 1-cocycle of Γ_k with coefficients in $C(\overline{k})$ defined by our trivialization of Y is a continuous map $g \mapsto c_g$ such that for any $g \in \Gamma_k$ we have ${}^g \overline{y}_0 = \overline{y}_0 c_g$. Hence we have ${}^g(\overline{y}_0 c) = \overline{y}_0 c_g \cdot {}^g c$ for any $c \in C(\overline{k})$. This means that ν represents Y as $\overline{C}$ with the twisted action of Γ_k given by $g(c) = c_g \cdot {}^g c$ (the usual action on $C(\overline{k})$ followed by the left translation by c_g). Let $b_g \in B(\overline{k})$ be any lifting of c_g such that $g \mapsto b_g$ is continuous. Then the map $z_g(b) = b_g \cdot {}^g b$, $b \in B(\overline{k})$, is a semilinear Y-automorphism of $\overline{B}$ which is a lifting of $g \in \Gamma_k$ to $\mathrm{Aut}(\overline{B}/Y)$. Hence (2.35) is right exact. The fact that A is central in B implies that for $a \in A(\overline{k})$ we have $z_g a_l z_g^{-1}(b) = {}^g a \cdot b$, where a_l denotes the left translation by a. This shows that the induced Γ_k-module structure on $A(\overline{k})$ is its usual Γ_k-module structure. The class of (2.35) can be represented by the 2-cocycle $z_g z_h z_{gh}^{-1}$. It is an immediate computation that it acts as the left translation by $b_g {}^g b_h b_{gh}^{-1}$. This is precisely the 2-cocycle $\partial(c_g)$. QED

Proof of Theorem 2.4.1 (concluded). To prove the claim before Lemma 2.4.3 we can replace U by Y. Indeed, we have $[*_U] = q^*[*_Y]$. On the other hand, the map $\alpha_n : \pi_1^{ab}(\overline{U}) \to R(\overline{k})[n]$ is the composition of $q_* : \pi_1^{ab}(\overline{U}) \to \pi_1^{ab}(\overline{Y})$ with $\pi_1^{ab}(\overline{Y}) \to \pi_1^{ab}(\overline{Y})/n = R(\overline{k})[n]$. The last equality is a canonical isomorphism by the Kummer sequence and the fact that $Pic(\overline{Y}) = 0$. It is clear that $\overline{Z}_n = \overline{U} \times_{\overline{Y}} \overline{Y}_n$, where $\overline{Y}_n$ is the unramified covering of $\overline{Y}$ corresponding to the surjection $\pi_1(\overline{Y}, \overline{y}) \to \pi_1^{ab}(\overline{Y})/n$. Now the desired identification of $\partial_n([*_Y])$ with the class of the extension $1 \to Aut(\overline{Y}_n/\overline{Y}) \to Aut(\overline{Y}_n/Y) \to \Gamma_k \to 1$ is a particular case of Lemma 2.4.5 applied to the central extension $1 \to R[n] \to R \to R \to 1$. QED

Exercises

1. Let X be a connected variety over an algebraically closed field k, and G an algebraic k-group. Prove that an X-torsor under G is not connected if and only if it can be obtained as a push-forward of an X-torsor under a closed subgroup of G of finite index.

2. Let X be a connected variety over an algebraically closed field k, and S be a k-group of multiplicative type. Prove that an X-torsor under S of type $\lambda : \hat{S} \to Pic(\overline{X})$ is not connected if and only if $Ker(\lambda)_{tors} \neq 0$.

3. Consider the quasi-affine surface $U \subset \mathbf{A}_k^3$ defined by

$$y^2 - bz^2 = aP(x) \neq 0,$$

where $a, b \in k^*$, and $P(x) \in k[x]$ is a separable monic polynomial. Let X be a smooth and geometrically integral compactification of U. Prove that X-torsors of type λ defined in Corollary 2.3.10 are universal. Show that they exist if and only if a is a product of a norm of the k-algebra $k(\sqrt{b})$ and the k-algebra $k[x]/(P(x))$.

4. Consider the quasi-affine curve $U \subset \mathbf{A}_k^2$ defined by

$$y^2 = aP(x) \neq 0,$$

where $a \in k^*$, and $P(x) \in k[x]$ is a separable monic polynomial. Let X be the smooth and proper hyperelliptic curve containing U. Prove that X-torsors of type λ defined in Corollary 2.3.10 exist if and only if a is a product of a square in k and a norm of the k-algebra $k[x]/(P(x))$.

5. Let X be a k-variety such that $\overline{k}[X]^* = \overline{k}^*$ and $Pic(\overline{X})$ is torsion free. Let Y/X be a universal torsor, and $Y \subset Y^c$ be a smooth compactification. Deduce from the exact sequence (2.9) that $Pic(\overline{Y}) = 0$, and then show that $Pic(\overline{Y^c})$ has a Γ_k-invariant base. Conclude that $\mathrm{Br}_1(Y^c) = \mathrm{Br}_0(Y^c)$. Prove that the partition of $Y^c(k)$ defined by a universal Y^c-torsor consists of the one set $X(k)$.

Comments

The material of the first two sections is mostly standard; needless to say, we owe it to a large extent to the immense legacy of Grothendieck and his school. Our main sources were [Milne, EC] and [Serre, CG].

Most of the third section is taken from [CS87a]. The theory of torsors under groups of multiplicative type was built by Colliot-Thélène and Sansuc who realized their importance for the systematic theory of descent. In a series of notes summarized in [CS80] and mainly in [CS87a], they introduced the concept of universal torsors, which found many applications, some of them hardly mentioned in this book (*R*-equivalence, 0-cycles, points of bounded height, counter-examples to the Zariski conjecture, etc.). Among the other things they proved that there is no 'second descent' on smooth compactifications of universal torsors over rational varieties, in the sense that universal torsors 'of second order' give no information about rational points (cf. Exercise 5 above).

Proposition 2.3.11, which establishes a link between the theory of Colliot-Thélène and Sansuc and the more common Leray spectral sequence, is taken from [HS], as well as all the material of the last section of this chapter. We refer to [HS] for the definition of a more general obstruction to the existence of rational points, which encompasses both the fundamental obstruction and Springer's obstruction on homogeneous spaces.

The fundamental obstruction to the existence of rational points appears to be a tiny fragment of 'the anabelian dream' of Grothendieck [Grothendieck 84]. We do not venture to speculate on the possible significance of this impressive edifice under construction to the study of rational points. A helpful short exposition of the current state of affairs can be found in T. Szamuely's survey in *Courbes semi-stables et groupe fondamental en géométrie algébrique*, Progress in Math. **187**, Birkhäuser, 2000. A major recent advancement is the work of S. Mochizuki, *The local pro-p anabelian geometry of curves*, Inv. Math. **138** (1999) 319–423.

3

Examples of torsors

The point of view of this chapter is that a torsor can be considered as the morphism of passing to the quotient by a freely acting algebraic group. To make this statement precise we evoke the basics of geometric invariant theory, and then treat in detail the example of a maximal torus of $PGL(5)$ acting on the Grassmannian $G(3,5)$. This leads to classification of Del Pezzo surfaces of degree 5. After a discussion of properties of torsors related to central extensions of algebraic groups, we describe explicit 2- and 4-descent on elliptic curves. Our intention in this chapter is to demonstrate in the examples the rôle played by the general concepts such as the type of a torsor, universal torsors, and so on.

3.1 Torsors in geometric invariant theory

Suppose that an algebraic k-group G acts on a k-variety Y. The following definition describes what could reasonably be called 'the quotient variety Y/G'.

Definition 3.1.1 ([Mumford, GIT], Def. 0.6) *The morphism* $\phi : Y \to X$ *is called a* **geometric quotient** *of* X *by* G *if*

(i) the action of G *preserves the fibres of* ϕ,

(ii) every geometric fibre of ϕ *is an orbit of a geometric point,*

(iii) ϕ *is universally open (for any base change* T/X *a subset* $U \subset T$ *is open if and only if* $U \times_T Y_T$ *is open in* Y_T*), and*

(iv) the structure sheaf $\mathcal{O}_X$ *is the* G-invariant subsheaf of $\phi_*(\mathcal{O}_Y)$.

This definition can be somewhat simplified. Suppose that $char(k) = 0$, X and Y are irreducible and normal, and ϕ is dominant; then *(iii)* and *(iv)* are automatically satisfied ([Mumford, GIT], Prop. 0.2). A geometric

42

quotient, when it exists, is unique up to isomorphism ([Mumford, GIT], Prop. 0.1).

The connection with torsors is the following: *when the action of G is free, the geometric quotient* $\phi : Y \to X$ *is an X-torsor under G* ([Mumford, GIT], Prop. 0.9). By definition, an action of G on Y is free if the map $G \times_k Y \to Y \times_k Y$ given by $(s, y) \mapsto (sy, y)$ is a closed immersion. (This condition is in general stronger than the triviality of the stabilizers of geometric points; however if Y and G are affine, and the characteristic is 0, then these are equivalent by Luna's étale slice theorem, see [Mumford, GIT], p. 199.)

The geometric invariant theory describes a general method of constructing geometric quotients of linear groups acting on smooth proper varieties. Let us recall the following important definition.

Definition 3.1.2 ([Mumford, GIT], Def. 1.7) *Let G be a reductive algebraic k-group; then*

*(i) a point $y \in Y(\overline{k})$ is **pre-stable** if y is contained in an invariant affine open subset $U \subset Y$ such that the G-orbits of all the $\overline{k}$-points of U are closed.*

Now let L be a G-linearized invertible sheaf on Y; then

*(ii) a point $y \in Y(\overline{k})$ is **semi-stable** with respect to L if there exists an invariant section $s \in H^0(Y, L^n)$ for some n, such that the subset $U_s \subset Y$ given by $s \neq 0$ is affine, and $y \in U_s(\overline{k})$,*

*(iii) a point $y \in Y(\overline{k})$ is **stable** with respect to L if there exists an invariant section $s \in H^0(Y, L^n)$ for some n, such that the subset $U_s \subset Y$ given by $s \neq 0$ is affine, $y \in U_s(\overline{k})$, and the G-orbits of all the $\overline{k}$-points of U_s are closed.*

The sets of pre-stable, semi-stable and stable points are open subsets of Y.

In this book we shall only consider the following 'ideal' situation. Let Y be a smooth, proper and geometrically irreducible k-variety, G a connected reductive k-group acting on Y, and L a G-linearized ample invertible sheaf on Y. Let Y^s (resp. Y^{ss}) be the set of stable (resp. of semi-stable) points. Then there is a geometric quotient $X = Y^s/G$, the morphism $Y^s \to X$ is affine, the variety X is quasi-projective and smooth ([Mumford, GIT], I.4). This leads to many classical examples of torsors. In this situation there is also an affine morphism $Y^{ss} \to X'$ to a (possibly singular) proper variety X', whose fibres are preserved by G. Actually,

$X' = Proj((\bigoplus_{i=0}^{\infty} H^0(Y, L^n))^G)$. Thus X is equipped with a natural compactification X'.

Example.

The simplest example of a torsor is, probably, the affine cone of a smooth projective variety X. The type of this torsor under $\mathbf{G}_m$ is (up to sign) the homomorphism sending $1 \in \mathbf{Z}$ to $[H \cap X] \in Pic(X)$, where H is a hyperplane.

Let us consider a more sophisticated example.

Let $G(m, n)$ be the Grassmannian variety of m-dimensional subspaces of the vector space k^n with a basis $\{e_1, \ldots, e_n\}$. Let $S \subset SL(n)$ be the diagonal torus. The stable points of $G(m, n)$ with respect to the action of S and the S-linearized sheaf $\mathcal{O}(1)$ correspond to the subspaces $V \subset k^n$ such that $dim(V \cap \langle e_{i_1}, \ldots, e_{i_p} \rangle) < (m/n)p$ (see [Mumford, GIT], the proof of Prop. 4.3). The quotient $X = G(m, n)^s/S$ is a smooth k-variety, which is projective if $(m, n) = 1$. This construction also produces an X-torsor under a torus which is the image D of S in $PGL(n)$. This torsor is closely related to the torsors on X. Indeed, the exact sequence (2.9) in this case is just

$$0 \to \hat{D} \to Pic(\overline{X}) \to \mathbf{Z} \to 0$$

since the geometric Picard group of $G(m, n)^s$ is $\mathbf{Z}$. In the appendix to this section we explore this construction for $n = 5$ and $m = 3$ in more detail.

We shall also see many examples when a finite group G acts freely on a quasi-projective k-variety Y. Then the quotient variety $X = Y/G$ always exists, and $Y \to X$ is étale (this is proved, for example, in [Mumford, AV], Thm. II.7.1). Then Y is an X-torsor under G. An example of such situation is when Y is a k-torsor under a connected k-group H, and G is a finite subgroup of H.

Appendix. Classification of Del Pezzo surfaces of degree 5. A well known observation is that $\mathbf{P}_k^2$ with four points in general position blown up is the moduli space $\overline{M}_{0,5}$ of stable curves of genus 0 with five marked points. The open set complementary to the six lines passing through the pairs of these four points is the moduli space $M_{0,5}$ of projective lines with five marked points. Grothendieck writes in his "Esquisse d'un programme" (p. 6) [Grothendieck, 84]:

J'ai commencé à regarder $M_{0,5}$ à des moments perdus, c'est un véritable joyau, d'une géométrie très riche étroitement liée à celle de l'icosaèdre.

The icosahedron will not appear in the sequel, but the symmetric group on five elements Σ_5 will play a crucial rôle. Note, however, that the icosahedron is indeed very useful in describing the intersection graph of the ten lines of $\mathbf{P}_k^2$ with four points in general position blown up, that is, four inverse images of the blown-up points and six lines passing through pairs of them (see the exercise at the end of this chapter).

A Del Pezzo surface of degree 5 over k is by definition a $(\overline{k}/k)$-form of $\mathbf{P}_k^2$ with four points in general position blown up. (The anticanonical class of such a surface is ample, and embeds it into $\mathbf{P}_k^5$ as a surface of degree 5.) The number 4 is a critical value: for $n \leq 4$ the group $PGL(3)$ acts transitively on n-tuples of k-points in $\mathbf{P}_k^2$ (in general position), so that the surfaces obtained by blowing up these points are isomorphic over $\overline{k}$. Hence the moduli space is a point. On the other hand, for $n \geq 4$ the automorphism group of the blowing up of $\mathbf{P}_k^2$ in n k-points in general position is a finite group (a subgroup of the Weyl group of the root system E_n, where by definition $E_4 = A_4$ and $E_5 = D_5$; see [Manin, CF], Ch. 4). This implies that Del Pezzo surfaces of degree 5 are classified by the elements of the first Galois cohomology set with coefficients in a finite group. We have the following more precise result.

Let Σ_n be the group of permutations of $\{1, 2, \ldots, n\}$. Note that Σ_5 is the Weyl group $W(A_4)$.

Theorem 3.1.3 *There are natural bijections between the following two sets:*

(i) the set of isomorphism classes of Del Pezzo surfaces of degree 5 over k,

(ii) the set of continuous homomorphisms of Γ_k to Σ_5 (equipped with discrete topology) considered up to conjugations in Σ_5 (equivalently, the set $H^1(k, \Sigma_5)$ with trivial action of Γ_k on Σ_5).

Recall that a *separable* (or *étale*) *k-algebra* L is by definition a commutative k-algebra isomorphic to a direct sum of finite separable extensions of k. The dimension of L as a vector space over k is called the *degree* of L. The separable k-algebras of degree d are the $(\overline{k}/k)$-forms of the commutative algebra k^d with coordinate-wise addition and multiplication. The automorphism group of this k-algebra is Σ_d. Hence the separable k-algebras of degree d are classified up to isomorphism by the elements of the Galois cohomology set $H^1(k, \Sigma_d)$ (with the trivial action of Γ_k on Σ_d).

Let $G(m, L)$ be the Grassmannian of m-dimensional subspaces of $L \times_k \overline{k}$. Define a k-torus S_L by the property that for a field extension K/k the group

$S_L(K)$ is the set of elements of $(L \times_k K)^*$ of norm 1. It is clear that S_L is a maximal torus in $SL(5)$.

Theorem 3.1.4 *Let X be a Del Pezzo surface of degree 5 over k. Then there exists a separable k-algebra L of degree 5, such that X is the geometric quotient of the set of stable points of $G(3, L)$ with respect to the natural action of S_L and the S_L-linearized sheaf $\mathcal{O}(1)$.*

Corollary 3.1.5 (Enriques, Swinnerton-Dyer) *Every Del Pezzo surface X of degree 5 defined over k has a k-point.*

Proof. By Theorem 3.1.4 we have a rational map $G(3, L) \simeq G(3, 5) - - >$ X. The statement of corollary is then obvious if k is infinite because then k-points are dense in $G(3, 5)$. In general, a lemma of Lang–Nishimura ([Lang 54], [N]) asserts that if we have a rational map $X' - - > X$, where X and X' are integral, X is proper, and X' has a smooth k-point, then X has a k-point. QED

The heart of the proof is the following lemma.

Lemma 3.1.6 *The geometric quotient of the set of stable points of $G(3, 5)$ with respect to the action of a split (say diagonal) maximal torus $S \subset SL(5)$, and the S-linearized sheaf $\mathcal{O}(1)$, is isomorphic to the blowing up of $\mathbf{P}_k^2$ in four k-points in general position.*

This clearly implies that the 'quotient' of $G(3, 5)$ by the action of any maximal torus of $SL(5)$ is a Del Pezzo surface of degree 5.

Proof of lemma. More generally, let $T \simeq \mathbf{G}_m^n$ be a split torus in $GL(n)$. The torus $S = T \cap SL(n)$ acts on $G(m, n)$. We can consider the sets of *stable* and *semi-stable* points of $G(m, n)$ with respect to the S-linearized ample sheaf $\mathcal{O}(1)$. Let us describe these sets.

Let $GL(n) \to GL(V)$ be the natural n-dimensional representation of $GL(n)$, and let $V = \bigoplus_{i=1}^n V_i$ be the decomposition into the direct sum of 1-dimensional T-invariant subspaces. If $I \subset \{1, 2, \dots, n\}$ we denote by V_I the direct sum of V_i for $i \in I$. One computes ([Mumford, GIT], the proof of Prop. 4.3) that an m-dimensional subspace $W \subset V$ defines a stable (resp. semi-stable) point of $G(m, n)$ if and only if $dim(W \cap V_I) < \frac{m}{n}|I|$ (resp. $dim(W \cap V_I) \le \frac{m}{n}|I|$) for all proper subsets $I \subset \{1, 2, \dots, n\}$. Let us choose $e_i \in V_i$, $e_i \ne 0$. This allows us to identify Σ_n with the subgroup of $GL(n)$ consisting of permutational $n \times n$ matrices (that is, $A = (a_{ij})$ such that $a_{ij} = \delta_{i,\pi(i)}$ where $\pi \in \Sigma_n$). Then Σ_n normalizes T and hence acts on the quotient of the set of stable points $Y = G(m,n)^s/T$ (resp. of semi-stable points $Y' = G(m,n)^{ss}/T$). Recall that according to the basic results of geometric invariant theory Y is smooth and Y' is proper. When m and n are coprime Y coincides with Y'.

Let D be the image of T in $PGL(n)$. We claim that $G(m,n)^s \to Y$ is a torsor under D. To prove this we cover Y by open sets over which this projection becomes a direct product (we follow [Mumford, GIT], 3.3). Suppose a point of $G(m,n)$ corresponds to the subspace generated by the rows of an $m \times n$ matrix M. If $I \subset \{1,2,\dots,n\}$, $|I| = m$, then let x_I be the determinant of the square submatrix of M formed by columns corresponding to the elements of I (these are Plücker coordinates). Let Z_I be the closed subset $x_I = 0$. The open sets U_I given by $x_I \neq 0$ form an open covering of $G(m,n)$. Assume for simplicity that $I = \{1,2,\dots,m\}$. Define an R-partition of $\{1,2,\dots,m\}$ as an ordered set of subsets $E_1,\dots,E_{n-m}$ which cover $\{1,2,\dots,m\}$, and such that

$$|E_i \cap (E_1 \cup \dots \cup E_{i-1})| = 1$$

for $i = 2,\dots,n-m$. Let E be an R-partition. We associate to E a dense open set $U_E \subset U_I$ given by $x_J \neq 0$ for all possible sets $J = I \cup \{m+j\} \setminus \{i\}$ where $i \in E_j$. One checks similarly to [Mumford, GIT] that U_E is affine, and moreover

$$U_E = D \times \mathbf{A}_k^{(m-1)(n-m-1)}.$$

A partition (in the usual sense) $\{1,\dots,n\} = P_1 \cup \dots \cup P_k$ and a decomposition $m_1 + \dots + m_k = m$ such that $0 < m_i \leq |P_i|$ define an obvious embedding of $G(m_1,|P_1|) \times \dots \times G(m_k,|P_k|)$ into $G(m,n)$ (here $G(m_i,|P_i|)$ is the set of m_i-dimensional subspaces of V_{P_i}). Let $U \subset G(m,n)$ be the complement to the union of images of all such embeddings. (This is the set of *pre-stable* points of the action of D, also characterized by the property that the geometric stabilizers are trivial; see [Mumford, GIT].) One checks that the union of the sets U_E for all possible permutations of indices is U. It is obvious that $G(m,n)^s \subset U$. Let $U_I^s = U_I \cap G(m,n)^s$, $U_E^s = U_E \cap G(m,n)^s$. Permuting the indices we get a D-invariant open covering of $G(m,n)^s$ with the property that restricting the morphism $G(m,n)^s \to Y$ to an open set of this covering we get a direct product of D and the quotient of this open set by D. This gives an explicit proof of the fact that $G(m,n)^s \to Y$ is a torsor under D.

In the case of interest here $n = 5$ and $m = 3$, and the stability condition can be easily seen to be equivalent to the following: if a subspace $U \subset V$ is generated by the rows of a 3×5 matrix M, then

(1) no column is 0,

(2) every two columns are linearly independent

(3) every four columns generate a 3-dimensional vector space.

To prove the lemma we let $Y_0 = \cap_I U_I^s/D$ where I ranges over all three-element subsets of $\{1, 2, 3, 4\}$. Points of Y_0 bijectively correspond to matrices M whose columns are (1,0,0), (0,1,0), (0,0,1), (1,1,1) and (x, y, z), where x, y, z are defined up to a common multiple, and (x, y, z) is not proportional to any of the first four columns. In other words, Y_0 is isomorphic to $\mathbf{P}_k^2$ without four k-points in general position. Since Y is a smooth, proper, geometrically irreducible surface, this isomorphism extends to a birational morphism $Y \to \mathbf{P}_k^2$. We now prove that the inverse of this morphism is the blowing up of these four points. Note that $Y \setminus Y_0$ is the union of four geometrically irreducible subsets Z_I^s/D, where I ranges over all three-element subsets of $\{1, 2, 3, 4\}$. One easily checks that no two of them intersect. Since any birational morphism of smooth and proper surfaces factors into a sequence of blowings up and their inverses, we conclude that each Z_I^s/D must be a projective line (since it is geometrically irreducible). This proves that Y is a Del Pezzo surface of degree 5. The lemma is proved. QED

As a corollary of the proof we get that the closed subsets Z_I^s/D where $I \subset \{1, 2, 3, 4, 5\}$, $|I| = 3$, are the exceptional curves of the first kind. It is well known [Manin, CF] that there are precisely ten such curves on Y, hence all of them are obtained in this way.

From now on we let Y be the blowing up of $\mathbf{P}_k^2$ in four k-points in general position, and X be an arbitrary $(\bar{k}/k)$-form of Y.

Lemma 3.1.7 *We have* $Aut(Y) = \Sigma_5$; *the action of* Γ_k *on this group is trivial.*

Proof. The action of $Aut(Y)$ on $Pic(\overline{Y})$ defines a homomorphism

$$\nu : Aut(Y) \to Aut(Pic(\overline{Y})).$$

Observe that $Ker(\nu) = 1$. (Indeed, suppose that $\alpha \in Aut(Y)$ fixes the classes of four disjoint exceptional curves on Y. Since every exceptional curve is the only effective divisor in its linear equivalence class, α leaves these four curves invariant. Then α defines an automorphism of $\mathbf{P}_k^2$ which fixes four points in general position, thus $\alpha = 1$.) This action leaves invariant the canonical class K_Y and the intersection form $(\cdot, \cdot)$. It is known ([Manin, CF], IV.1) that the subgroup of $Aut(Pic(\overline{Y}))$ leaving invariant K_Y and $(\cdot, \cdot)$ is isomorphic to Σ_5 (the Weyl group of the root system A_4).

On the other hand, we have seen that Σ_5 acts on Y (since it normalizes the split maximal torus in $GL(5)$). Thus $\nu(Aut(Y))$ contains Σ_5, and hence must be equal to it. QED

Theorem 3.1.3 immediately follows from this lemma.

Proof of Theorem 3.1.4. We already know that Y is the geometric quotient of $G(3, k^5)^s$ by the action of the diagonal torus $S \subset SL(5)$. An arbitrary Del Pezzo surface X of degree 5 is a $(\overline{k}/k)$-form of Y. Let ϕ be a continuous homomorphism $\Gamma_k \to \Sigma_5$ such that $X = {}_\phi Y$ is the twist of Y by ϕ. The group Σ_5, realized as the group of permutational matrices, acts on $\overline{k}^5$ by permutations, and on S and $SL(5)$ by conjugations, and it acts on $G(3,5)$ as a subgroup of $GL(5)$. We can consider the twisted forms of these objects: ${}_\phi(\dot{k}^5) = L$, ${}_\phi S = S_L$, and ${}_\phi G(3,5) = G(3,L)$. Then X is the geometric quotient of ${}_\phi(G(3,5)^s) = G(3,L)^s$ by S_L. QED

3.2 Homogeneous spaces and central extensions

Perhaps the simplest and the most fundamental example of a torsor is provided by the morphism $B \to X = B/A$, where A is a closed algebraic k-subgroup of an algebraic k-group B.

We make the assumption that *the quotient $X = B/A$ exists*, that is, there is a morphism of k-varieties $B \to X$ whose $\overline{k}$-fibres are the cosets of $\overline{A}$ in $\overline{B}$. This assumption is satisfied if (1) B is affine, then the quotient X is a quasi-projective homogeneous space of B (a consequence of Chevalley's theorem; see [Springer 81], 5.1.4, 5.2.2), or (2) if A is finite.

Therefore B is an X-torsor under A. The k-fibres of the natural map $B \to X$ are k-torsors under A acting on the right.

Recall the following exact sequence of pointed sets ([Serre, CG], Prop. I.36):

$$1 \to A(k) \to B(k) \to X(k) \xrightarrow{\delta} H^1(k, A) \to H^1(k, B). \qquad (3.1)$$

The map δ sends $P \in X(k)$ represented by $b \in B(\overline{k})$ to the 1-cocycle of Γ_k given by $g \mapsto a_g$ where $ba_g = {}^g b$ (*ibidem*). Thus the map θ_B associating to P the class of the fibre of $B \to X$ at P coincides with δ. This gives a transparent description of the partition of $X(k)$ defined by the torsor $B \to X$.

Remark. The class of this torsor in the Čech cohomology group $\check{H}^1(X, A)$ can be seen as the image of $Id \in Mor_k(X, X)$ in $\check{H}^1(X, A)$ in the following exact sequence of pointed sets ([Giraud], III.3.2.2):

$$1 \; \to \; Mor_k(X, A) \; \to \; Mor_k(X, B) \; \to \; Mor_k(X, X)$$
$$\to \; \check{H}^1(X, A) \; \to \; \check{H}^1(X, B) \tag{3.2}$$

The homogeneous space X just considered always has a k-point inherited from B. On twisting by Galois descent as in Section 2.1, we arrive at a more general class of homogeneous spaces, in general without a k-point. Let P be a k-torsor under B. Consider the twist $_PX = P \times^B X$ of X, or, in other words, the quotient scheme P/A. One checks immediately ([Serre, CG], Prop. I.37) that $_PX(k) \neq \emptyset$ if and only if the class of P in $H^1(k, B)$ belongs to the image of $H^1(k, A)$. This means that we have an explicit criterion for the existence of k-points on such homogeneous spaces. In general, not every homogeneous space X under an algebraic k-group B is of the form $_P(B/A) = P/A$. A device that decides whether it is so or not is a (non-abelian) 2-cocycle (or a gerb) associated to the pair $(X$, the action of $B)$. We shall speak about gerbs in the last chapter.

The case of a normal subgroup.

When A is closed and normal in a k-group B, then $C = B/A$ is an algebraic k-group. (If B is affine, then C is also affine; [Springer 81], 5.2.5.) We have an extension of algebraic k-groups:

$$1 \to A \to B \to C \to 1. \tag{3.3}$$

In this case the above twisting construction produces k-torsors under C of the form P/A where P is a k-torsor under B. The exact sequence (3.1) can be extended one step further to the right:

$$1 \to A(k) \to B(k) \to C(k) \xrightarrow{\delta} H^1(k, A) \to H^1(k, B) \to H^1(k, C). \tag{3.4}$$

The class of P/A in $H^1(k, C)$ is the image of the class of P in $H^1(k, B)$.

Central extensions.

Let (3.3) be a *central* extension of k-groups, that is, A is contained in the centre of B. For central extensions the exact sequence of pointed sets (3.4) can be extended one more term to the right ([Serre, CG]):

$$1 \to A(k) \to B(k) \to C(k) \xrightarrow{\delta} H^1(k, A) \to H^1(k, B) \to H^1(k, C)$$
$$\xrightarrow{\delta} H^2(k, A) \tag{3.5}$$

(The last differential was defined by Grothendieck also for central extensions of sheaves of groups on a topological space; see [Grothendieck 57].)

The classical examples of central extensions of algebraic groups are the universal coverings of semisimple groups, isogenies of abelian varieties, and extensions of groups of multiplicative type.

Remark on extensions of groups of multiplicative type. When B is a group of multiplicative type, so are also A and C. The importance of the C-torsor B under A given by this extension stems from the fact that locally every torsor under a group of multiplicative type can be obtained as the pull-back of (3.3). More precisely, for any k-variety X and any X-torsor $Y \to X$ under A there exist a non-empty open subset $U \subset X$ and a map $U \to C$ for some extension (3.3) such that $Y_U = B \times_C U$; see Section 4.3.

We go back to central extensions of possibly non-abelian groups.

Lemma 3.2.1 *Let (3.3) be a central extension of algebraic k-groups, and let X be a right k-torsor under C. Let $\nu : \overline{C} \to \overline{X}$ be an isomorphism of right torsors under $\overline{C}$; then the class of $\overline{B} \to \overline{X}$ (the composite map) in $H^1(\overline{X}, \overline{A})$ is independent of ν, and is Γ_k-invariant.*

Proof. Let $\nu' : \overline{C} \to \overline{X}$ be another isomorphism of right torsors under $\overline{C}$. Then $\nu'\nu^{-1}$ is an automorphism of $\overline{X}$ as a right torsor under $\overline{C}$, hence this is the left translation by an element of $C(\overline{k})$. Lift it to an element of $B(\overline{k})$. The left translation by it is a lifting of $\nu'\nu^{-1}$ to an isomorphism of the corresponding $\overline{X}$-torsors under $\overline{A}$. This proves that this isomorphism class in $H^1(\overline{X}, \overline{A})$ is well defined. It is clear that the torsor $\overline{B} \to \overline{C}$ and its conjugates are isomorphic as torsors under $\overline{A}$ (since these varieties and morphisms are defined over k). The conjugate of ν is an isomorphism of $\overline{C}$ and $\overline{X}$ as right torsors under $\overline{C}$, and we use the same argument as before to conclude that the conjugate of our $\overline{X}$-torsor under $\overline{A}$ is isomorphic to it. QED

We denote this class by $\tau \in H^0(k, H^1(\overline{X}, \overline{A}))$. When A is of multiplicative type this group coincides with $Hom_k(\hat{A}, Pic(\overline{X}))$ (by Proposition 2.3.11). The lemma could be seen as a particular case of [Giraud], V.3.2.9.

We have the following useful property.

Proposition 3.2.2 ([Giraud], **V.3.2.9**) *In the notation of Lemma 3.2.1 let us further assume that $H^0(\overline{X}, \overline{A}) = \overline{A}(\overline{k})$, for instance, A is finite and C is geometrically connected and geometrically reduced, or A is of multiplicative type and $\overline{k}[C]^* = \overline{k}^*$, or A is affine and C is projective. Then we* *have*

$$\partial(\tau) = \delta([X]),$$

where $\partial : H^0(k, R^1 p_* A) \to H^2(k, A)$ *is a differential from the Leray spectral sequence (2.14), and* $\delta : H^1(k, C) \to H^2(k, A)$ *is from (3.5).*

The following proposition has many applications.

Proposition 3.2.3 ('Lifting property' of torsors) *Let (3.3) be a central extension of algebraic* k-*groups such that* A *is a* k-*group of multiplicative type. Assume that* $\overline{k}[C]^*/\overline{k}^* = 1$. *Let* X *be a* k-*torsor under* C. *Then any* X-*torsor* $f : Y \to X$ *under* A *of type* τ *(defined after the proof of Lemma 3.2.1) can be endowed with the structure of a* k-*torsor under* B *which extends the action of* A *on* Y.

Proof. By the definition of a torsor there exists a morphism $m : X \times_k C \to X$ such that $(p_1, m) : X \times_k C \to X \times_k X$ is an isomorphism (p_i is the projection to the i-th factor). We have to show that there is a morphism $m' : Y \times_k B \to Y$ compatible with the obvious action of $A \times_k A$ ((a_1, a_2) acts on $Y \times_k B$ by sending (y, b) to (ya_1, ba_2), and on Y by sending y to ya_1a_2), such that $(p_1, m') : Y \times_k B \to Y \times_k Y$ is an isomorphism. In other words, the following diagram should be commutative and compatible with the action of $A \times_k A$ (the action is trivial on X and $X \times_k C$):

$$
\begin{array}{ccc}
Y \times_k B & \xrightarrow{\ m'\ } & Y \\
\downarrow & & \downarrow \\
X \times_k C & \xrightarrow{\ m\ } & X
\end{array}
\qquad (3.6)
$$

Here the left vertical map is given by the pair of maps $f : Y \to X$ and $B \to C$. We claim that the existence of (3.6) commutative and compatible with the action of $A \times_k A$ is equivalent to the property that the pull-back of Y to $X \times_k C$ is isomorphic to the contracted product of the pull-backs $p_1^*(Y)$ and $p_2^*(B)$ over $X \times_k C$:

$$m^*(Y) = p_1^*(Y) \times^A p_2^*(B). \qquad (3.7)$$

Recall that the contracted product is the quotient of $Y \times_k B$ by the action of A given by $(y, b) \mapsto (ya, ba)$. Let us denote it by Z. By the definition of the group law on the set of isomorphism classes of $(X \times_k C)$-torsors under A, $Z \to X \times_k C$ is a torsor under A such that $[Z] = p_1^*[Y].p_2^*[B] \in H^1(X \times_k C, A)$.

Suppose we have (3.6). This implies that the following diagram commutes and is compatible with the action of A:

$$
\begin{array}{ccc}
Z & \longrightarrow & Y \\
\downarrow & & \downarrow \\
X \times_k C & \xrightarrow{\ m\ } & X
\end{array}
\qquad (3.8)
$$

Then Z factors through the fibre product of Y and $X \times_k C$ over X, hence we get a morphism $Z \to m^*(Y)$ of $(X \times_k C)$-torsors under A. Any such morphism is an isomorphism, hence we get (3.7).

Conversely, (3.7) implies that there exists a morphism $m_1 : Z \to Y$ making (3.8) commutative and compatible with the action of A. Define m' as the composition of $Y \times_k B \to Z$ with m_1. The corresponding properties for (3.6) follow from this.

Let us establish (3.7). To check that two torsors under a group of multiplicative type are isomorphic, it suffices by (2.23) to check that they have identical types, and coincide on some subvariety $X' \subset X$ such that $H^0(\overline{X}', \mathbf{G}_m) = \overline{k}^*$. To check the type we can work over $\overline{k}$. Since (3.3) is an extension of algebraic groups, there is a group structure morphism $\mu' : B \times_k B \to B$. It is compatible with the action of $A \times_k A$. We have (3.6) with B in place of Y. The preceding analysis now tells us that this implies

$$
\mu^*(B) = p_1^*(B) \times^A p_2^*(B), \qquad (3.9)
$$

where $\mu : C \times_k C \to C$ is the group structure morphism. On passing to $\overline{k}$ the morphism m becomes μ. Thus over $\overline{k}$ the formula (3.9) is equivalent to $m^*(\overline{Y}) = p_1^*(\overline{Y}) \times^{\overline{A}} p_2^*(\overline{B})$. Hence $m^*(Y)$ and $p_1^*(Y) \times^A p_2^*(B)$ have the same type as $(X \times_k C)$-torsors. Finally, the restriction of any of these torsors to $X \times 1 \subset X \times_k C$ is just $Y \to X$. This proves (3.7).

To complete the proof we must show that $(p_1, m') : Y \times_k B \to Y \times_k Y$ is an isomorphism. In fact, we have a commutative diagram

$$
\begin{array}{ccc}
Y \times_k B & \xrightarrow{(p_1, m')} & Y \times_k Y \\
\downarrow & & \downarrow \\
X \times_k C & \xrightarrow{(p_1, m)} & X \times_k X
\end{array}
\qquad (3.10)
$$

The bottom arrow is an isomorphism. The upper arrow is compatible with the action of $A \times_k A$ with respect to the usual action on $Y \times_k B$, and the action on $Y \times_k Y$ given by $(y_1, y_2) \mapsto (y_1 a_1, y_2 a_1 a_2)$. Thus the upper arrow is a morphism of torsors under $A \times_k A$, and hence must be an isomorphism. QED

In fact, one can regard Proposition 3.2.2 as a corollary of Proposition

3.2.3 if one is allowed to use the language of gerbs. Indeed, Proposition 3.2.3 implies that the gerb of liftings of X_K to a K-torsor under B_K for all possible extensions K/k coincides with the gerb of X_K-torsors under A_K of type τ. Because of the condition that C is connected and $\overline{k}[C]^* = \overline{k}^*$, and hence $p_* p^* A = A$, the last gerb is also bound by the lien of A (recall that $p : X \to Spec(k)$ is the structure morphism). Hence the corresponding cohomology classes in $H^2(k, A)$ coincide.

The universal covering of a semisimple group as a universal torsor.

Let C be a semisimple k-group (in particular, it is geometrically connected). It is well known (cf. [PR], sect. II.2, Prop. 10) that there exists a semisimple k-group B such that $\overline{B}$ is simply connected (any connected étale covering is the identity), equipped with a surjective homomorphism $B \to C$, moreover, its kernel A is a finite *central* k-group subscheme of B. Let us assume that the characteristic of k is coprime with the torsion of A. Note that by Rosenlicht's lemma the property $\overline{k}[C]^* = \overline{k}^*$ is true for semisimple groups.

The k-group B is a C-torsor under A, and we have the 'lifting property' of torsors from Proposition 3.2.3. If X is a k-torsor under C, then X-torsors Y under A of type τ are universal torsors. To see this, it suffices to show that the second arrow in the exact sequence (2.4),

$$0 \to \hat{A}(\overline{k}) \to Pic(\overline{C}) \to Pic(\overline{B}),$$

is an isomorphism. It is well known that the Picard group of a simply connected semisimple group over $\overline{k}$ is zero (see, e.g., [Sansuc 81], 6.5, 6.9), hence $Pic(\overline{B}) = 0$. Thus *the universal torsors of type τ are precisely the liftings of X to a k-torsor under B.*

Recall that the elementary obstruction $e(X)$ (= the obstruction to the existence of universal X-torsors) is the inverse of the class of the 2-fold extension (2.18). For k-torsors X under semisimple groups the elementary obstruction has a very clear interpretation: it is just the second coboundary δ of (3.5) applied to the class of X in $H^1(k, C)$ (Proposition 3.2.2).

Isogenies of abelian varieties are another instance of the preceding proposition.

3.3 Torsors under abelian varieties

Let us start by recalling some well known facts about torsors on abelian varieties over an algebraically closed field of characteristic 0.

If C is an abelian variety, and A is a connected commutative *unipotent*

$\overline{k}$-group, then every C-torsor under A comes from an extension of commutative algebraic $\overline{k}$-groups

$$1 \to A \to B \to C \to 1, \tag{3.11}$$

and this defines a bijection between $Ext^1_{k\text{-groups}}(C, A)$ taken in the category of commutative algebraic $\overline{k}$-groups and $H^1(C, A)$ ([Serre, GA], VII. Thm. 8). When $A = \mathbf{G}_m$, then not every C-torsor under $\mathbf{G}_m$ can be obtained from an extension (3.11): this is the case if and only if the class of this torsor in $Pic(C) = H^1(C, \mathbf{G}_m)$ is algebraically equivalent to 0 ([Serre, GA], VII. Thm. 6).

The Serre–Lang theorem states that every étale covering of an abelian variety can be made into an isogeny (after an appropriate choice of base points). In particular, any torsor over an abelian variety C under a finite abelian group A comes from an extension of algebraic $\overline{k}$-groups (3.11) ([Grothendieck, SGA 1], XI.2, [Mumford, AV], IV.18, [Milne, EC], III.4.21). The type of this torsor is given by the natural isomorphism

$$\hat{A}(\overline{k}) \xrightarrow{\sim} Ker[Pic(\overline{C}) \to Pic(\overline{B})].$$

We go back to the arbitrary field k. Recall that to a smooth and projective variety X (possibly without a k-point) one associates its Albanese variety. It is an abelian variety over k, defined up to isomorphism by a certain universal property; see [Lang, AV], II.3. Its dual abelian variety parametrizes divisor classes algebraically equivalent to 0 modulo rational equivalence. The group of such divisor classes is denoted by $Pic^0(\overline{X})$.

If X is a variety over k such that $\overline{X}$ is isomorphic to an abelian variety, then it is easy to prove that X is a k-torsor under its own Albanese variety. (Let A be an Albanese variety of X. Choose a point $\overline{x}_0 \in X(\overline{k})$; then the $\overline{k}$-morphism $\overline{A} \to \overline{X}$ given by $a \mapsto a + \overline{x}_0$ descends to a k-morphism $Y \to X$ where Y is a k-torsor under A given by the cocycle $\gamma \mapsto {}^\gamma\overline{x}_0 - \overline{x}_0$.)

In what follows we assume that n is coprime with the characteristic of k. In the case of the endomorphism of an abelian variety $n : B \to B = C$ given by multiplication by n we endow B with the structure of a torsor over itself under $B[n] = A$. Its type is the natural injection $\lambda_n : B'[n](\overline{k}) \hookrightarrow B'(\overline{k})$ followed by the inclusion $B'(\overline{k}) = Pic^0(\overline{B}) \to Pic(\overline{B})$, where B' is the dual abelian variety of B. In the case of an algebraically closed field it is clear from Section 2.3 that $n : \overline{B} \to \overline{B}$ is the unique $\overline{B}$-torsor of type λ_n.

Definition 3.3.1 *Let B be an abelian variety over k. An n-**covering** of B is a pair (X, ψ), where X is a k-torsor under B, and $\psi : X \to B$ is a* *morphism such that*

$$\psi(\overline{b} \cdot \overline{x}) = n\overline{b} + \psi(\overline{x}) \qquad (3.12)$$

for any $\overline{b} \in B(\overline{k})$, $\overline{x} \in X(\overline{k})$.

Here are some of the most important properties of n-coverings.

Proposition 3.3.2 *(i) A k-torsor X under B can be endowed with the structure of an n-covering if and only if its class $[X]$ is in $H^1(k, B)[n]$.*

(ii) A morphism $\psi : X \to B$ can be endowed with the structure of an n-covering if and only if it is a $(\overline{k}/k)$-form of the multiplication by n, that is, a choice of the base point of $\overline{X}$ over $0 \in B(k)$ turns $\overline{X} \to \overline{B}$ into the multiplication by n.

(iii) Any n-covering is a B-torsor under $B[n]$, and can be obtained by twisting the torsor $n : B \to B$ under $B[n]$ by a cocycle from $H^1(k, B[n])$. The torsor under $B[n]$ corresponding to an n-covering (X, ψ) is recovered as $\psi^{-1}(0)$ where 0 is the neutral element of the group law on B.

Proof. (*i*) Let X be an n-covering of B, and let $n_*(X)$ be the push-forward of X with respect to the change of the structure group $n : B \to B$. By the definition of push-forward we have $n_*(X) = X \times_k^B B = (X \times_k B)/B$, where the action of B is given by $(\overline{x}, \overline{b}) \mapsto (-s\overline{x}, ns + \overline{b})$. Then $(Id, -\psi)(X) \subset X \times_k B$ is a k-orbit of B. It gives rise to a k-point on the quotient, which is therefore a trivial torsor. This implies that $n[X] = 0$. Conversely, if $n[X] = 0$, then $n_*(X) \simeq B$, and the canonical map $X \to n_*(X)$ is an n-covering.

(*ii*) It is clear from the definition that an n-covering is a form of the multiplication by n. Let us prove the converse. Let B_1 be the Albanese variety of X; then X is a k-torsor under B_1. We have a natural morphism of abelian varieties $\psi_* : B_1 \to B$ defined over k. It is the multiplication by n (we can check this over $\overline{k}$, where it immediately follows from our assumption). Hence $B_1 = B$. This shows that X is a k-torsor under B. Formula (3.12) can be checked over $\overline{k}$, where it follows from the assumption.

(*iii*) The map $n_*(X) \to B$ given by $(\overline{x}, \overline{b}) \mapsto \psi(\overline{x}) + \overline{b}$ is a unique isomorphism compatible with the action of B. It identifies $n_*(X)$ with B; then ψ is just the canonical morphism $X \to n_*(X) = B$. This is clearly a torsor under $B[n]$. Twisting it by the class of the zero fibre $\psi^{-1}(0)$ we find a k-point over 0, that is, we get the map $n : B \to B$. The proposition is proved. QED

Let m be also coprime with the characteristic of k.

Definition 3.3.3 *An nm-covering* $\psi' : X' \to B$ *is called a* **lifting** *of an n-covering* $\psi : X \to B$ *if* ψ' *factors through* ψ.

The nm-covering $\psi' : X' \to B$ is a lifting of the n-covering $\psi : X \to B$ if and only if the corresponding cocycles are related by the map $m_* : H^1(k, B[mn]) \to H^1(k, B[n])$. This implies that $m[X'] = [X] \in H^1(k, B)$.

The following proposition can be proved by the technique of Section 3.2, but there is a direct simple proof.

Proposition 3.3.4 *(a) Any n-covering of B is a B-torsor under $B[n]$ of type λ_n, and vice versa.*

(b) Any lifting $Y \to X$ of an n-covering $X \to B$ to an nm-covering $Y \to B$ is an X-torsor under $B[m]$ of type λ_m, and vice versa.

Proof. Part *(a)* is a particular case of *(b)*, so it is enough to prove *(b)*. The first assertion is clear. Let us prove the converse. Let $f : Y \to X$ be an X-torsor of type λ_m, where X is a k-torsor under B. Since over $\overline{k}$ the multiplication by m is the only $\overline{B}$-torsor of type λ_m, the morphism $\overline{Y} \to \overline{X}$ can be identified with the multiplication by m after an appropriate choice of the base points. In the same way as in the proof of the previous proposition one deduces from this that Y is also a k-torsor under B. The composition $Y \to X \to B$ is then a $(\overline{k}/k)$-form of the multiplication by mn. Now use *(ii)* of the previous proposition. QED

Proposition 3.3.5 *Let X be a k-torsor under an abelian variety B. Then an X-torsor of type λ_n exists if and only if the class $[X] \in H^1(k, B)$ is divisible by n. If this is the case, then the X-torsors $f : Y \to X$ of type λ_n are precisely the k-torsors Y under B such that $n[Y] = [X]$. Moreover, X is naturally isomorphic to $n_*(Y)$, and after this identification f becomes the canonical morphism $Y \to n_*(Y)$.*

Proof. Let $f : Y \to X$ be a torsor of type λ_n. Similar to the previous proofs one shows that $\overline{Y} \to \overline{X}$ can be identified with $m : B \to B$ after a choice of the base points such that $f(\overline{y}_0) = \overline{x}_0$, and that Y is also a k-torsor under B. The cocycle σ_g defining Y is given by $^g\overline{y}_0 = \overline{y}_0\sigma_g$, and the cocycle σ'_g defining X is given by $^g\overline{x}_0 = \overline{x}_0\sigma'_g$. Now the latter cocycle is the n-th power of the former. QED

The last two propositions are in fact equivalent since $H^1(k, B)$ is a torsion group. (There exists a finite extension K/k such that $X(K) \neq \emptyset$. Then the class $[X]$ is killed by $Cores_{K/k}Res_{K/k}$ which is multiplication by the degree $m = [K : k]$.)

Appendix. Equations for 2- and 4-coverings of elliptic curves.
Let the characteristic of k be different from 2. It is known how to give 2-
coverings and, sometimes, 4-coverings over elliptic curves by explicit equations. This old subject goes back to Fermat (see [Weil 83], Ch. II, App.
III). A much more general exposition explaining why the equations are such
as they are can be found in Section 4.4.

We start with a classical example of a 2-covering assuming for simplicity
that our elliptic curve E has its 2-division points defined over k. Then E
is given by

$$y^2 = (x - c_1)(x - c_2)(x - c_3), \qquad (3.13)$$

for some pairwise different constants $c_1, c_2, c_3 \in k$. Let $m_1, m_2, m_3 \in k^*$
be such that $m_1 m_2 m_3$ is a square in k^*. Consider the curve D in the
3-dimensional projective space over k given by the affine equations

$$X_i^2 = m_i(x - c_i), \quad i = 1, 2, 3, \qquad (3.14)$$

after eliminating x. This amounts to considering the projective equations

$$m_i^{-1} X_i^2 - m_j^{-1} X_j^2 = (c_j - c_i) X_0^2 \qquad (3.15)$$

for all $i \neq j$, $\{i, j\} \subset \{1, 2, 3\}$. One checks that D is smooth, hence
a curve of genus 1. The morphism $\psi : D \to E$ is defined by sending
(X_0, X_1, X_2, X_3) to $(X_1 X_2 X_3, x)$ where x is defined by (3.14). The map
ψ has degree 4, and one checks by local computations at the ramification
points $(x, y) = (c_i, 0)$ that ψ is étale. There are many ways to see that
$\psi : D \to E$ is a 2-covering. The simplest way is to notice that on E we
have $div(x - c_i) = 2(c_i, 0) - 2\infty$. It follows from (3.14) that $\psi^{-1}((c_i, 0) - \infty)$
is the divisor of the function X_i, that is, it is equivalent to 0 in $Pic(\overline{D})$.
Hence the kernel of $\psi^* : Pic(\overline{E}) \to Pic(\overline{D})$ contains $E[2] = Pic(\overline{E})[2]$.
Since the degree of ψ is 4, we must have $Ker(\psi^*) = E[2]$, thus the type of
ψ is indeed λ_2.

Another proof is to compute that the equation of E identifies with the
equation of the 'discriminantal curve' of the pencils of quadrics passing
through D. Recall the following well known geometric observation.

Let $Q(\mathbf{x})$ and $Q'(\mathbf{x})$ be quadratic forms in four variables, and let $\mathbf{x} = (x_0, x_1, x_2, x_3)$. Let $X \subset \mathbf{P}_k^3$ be the intersection of two quadrics given by
$Q(\mathbf{x}) = Q'(\mathbf{x}) = 0$. Let Y be the curve parametrizing families of lines on
the quadrics of the pencil spanned by Q and Q'. Then Y is canonically
identified with the degree 2 component $Pic^2(X)$ of the Picard scheme of
X: one associates to a family of lines on a quadric the divisor class of the
intersection of any line of this family with X. The curve Y comes with the

map $\xi : X \to Y = Pic^2(X)$ sending a point P to the divisor class of $2P$ (geometrically this is the family of lines containing the tangent to X at P). On the other hand, Y is isomorphic to the curve $\mu^2 = det(\lambda Q - Q')$ (not canonically, the two natural identifications differ in the sign of μ).

Thus if Y is isomorphic to an elliptic curve E, then ξ turns X into a 2-covering of E. Sometimes one can go one step further.

Proposition 3.3.6 (a) Let C be the curve of genus 1 given by the equation

$$y^2 = g(x),$$

where $g(x) = ax^4 + cx^2 + dx + e$ is a separable polynomial of degree 4 with coefficients in k. Then C can be equipped with the structure of a 2-covering $\psi : C \to E$ of the elliptic curve

$$E : \quad u^2 = v^3 - 27Iv - 27J,$$

where $I = 12ae + c^2$, $J = 72ace - 27ad^2 - 2c^3$, such that $\psi^{-1}(0) \subset C$ is given by $y = 0$.

(b) Let K be the algebra $k[x]/(ax^4 + cx^2 + dx + e)$, and let θ be the image of x in K. Suppose that the class of a in k^*/k^{*2} is represented by a norm $N_{K/k}(\epsilon)$ for some $\epsilon \in K$. Let $C' \subset \mathbf{P}_k^3$ be the intersection of two quadrics obtained by equating to 0 the coefficients of θ^2 and θ^3 in the following formula:

$$X - \theta Z = \epsilon(x_0 + x_1\theta + x_2\theta^2 + x_3\theta^3)^2.$$

Then C' can be equipped with the structure of a 4-covering of E which is a lifting of $\psi : C \to E$.

(c) If $C(k) = \emptyset$, then $[C']$ is of exact order 4 in the Weil–Châtelet group $H^1(k, E)$.

When $a = 1$ the passage from the equation of C to that of E is precisely the passage from a quartic polynomial to its cubic resolvent.

What this proposition describes is the following situation: C' is any 4-covering of E such that the corresponding 2-covering $C \to E$, $[C] = 2[C'] \in H^1(k, E)$, has a k-divisor of degree 2.

Proof of proposition. (a) The curve C is isomorphic to the intersection of the following two quadrics in $\mathbf{P}_k^3$:

$$Q = ut - x^2, \quad Q' = -y^2 + au^2 + cut + dxt + et^2.$$

One computes that

$$det(\lambda Q - Q') = (1/4)(\lambda^3 - 2c\lambda^2 + (c^2 - 4ae)\lambda + ad^2),$$

which gives the equation of E after an obvious change of variables. Thus $Y = Pic^2(C)$ of the above observation is just E. Let us choose the origin of the group law of E at $\lambda = \infty$ (this point corresponds to the hyperelliptic divisor on C). The map $\psi = \xi$ is a 2-covering, and it is easy to see that $\psi^{-1}(0)$ is as required.

(b) Let $\theta_i \in \overline{k}$ be the roots of the equation $ax^4 + cx^2 + dx + e = 0$. Let $\epsilon_i \in \overline{k}$ be the image of ϵ under the map $K \to \overline{k}$ which sends θ to θ_i. Consider the following two quadratic forms individually defined over k:

$$Q(\mathbf{x}) = \sum_{i=1}^{4} \epsilon_i \prod_{j \neq i} (\theta_i - \theta_j)^{-1} \Big(\sum_{j=0}^{3} x_j \theta_i^j \Big)^2,$$

$$Q'(\mathbf{x}) = \sum_{i=1}^{4} \epsilon_i \theta_i \prod_{j \neq i} (\theta_i - \theta_j)^{-1} \Big(\sum_{j=0}^{3} x_j \theta_i^j \Big)^2.$$

One computes that the quadrics given by $Q(\mathbf{x}) = 0$ and $Q'(\mathbf{x}) = 0$ contain C', and that we have

$$det(\lambda Q - Q') = N_{K/k}(\epsilon) \prod_{i=1,2,3,4} (\lambda - \theta_i).$$

This implies that the intersection of these two quadrics is a smooth and geometrically integral curve, hence C' is given by $Q(\mathbf{x}) = Q'(\mathbf{x}) = 0$. Thus $Y = Pic^2(C')$ is isomorphic to C, and after choosing one of two possible isomorphisms we have the map $\xi : C' \to C$. A structure of a 4-covering $\psi' : C' \to E$ is defined by the map sending a point P to the divisor class of $4P$, and identifying $Pic^4(C')$ with E by choosing the hyperplane section class as the origin of the group law. Then $\psi' = \psi \circ \xi$, so that ψ' is a lifting of ψ.

(c) Since ψ' is a lifting of ψ we have $2[C'] = [C] \in H^1(k, E)$. When C has no k-point, $[C] \neq 0$, and our statement is proved. QED

Exercise

Prove that the intersection graph of the ten lines on a Del Pezzo surface of degree 5 coincides with the quotient of the graph formed by vertices and edges of the icosahedron by the antipodal involution.

Comments

The fundamentals of geometric invariant theory are taken from [Mumford, GIT]. The appendix to the first section is a somewhat extended exposition of [S93], and Theorem 3.1.4 is a more precise variant of the main result of that paper. The classification of Del Pezzo surfaces of degree 5 undoubtedly can be carried over to the classification of the twisted forms of quotients of Grassmannians and more general homogeneous spaces. It was stated by Enriques more than 100 years ago that every Del Pezzo surface contains a k-rational point. This result is then used to prove the k-rationality of such surfaces. However, Enriques's proof was hard to follow. A new proof was given later by Swinnerton-Dyer [SD72], and a more recent one-page proof was found by Shepherd-Barron [SB92]. These proofs rely on specific geometric constructions. Our approach clarifies the structure of these surfaces and related torsors.

The key result of the second section is 'the lifting property' of torsors related to central extensions of algebraic groups (Proposition 3.2.3). Despite its technical appearance it is quite useful in applications. The theory of principal homogeneous spaces of abelian varieties is classical. The presentation in the last appendix in this chapter follows [S99], it was influenced by [Weil 83] and [MSS]. See also [Cassels 98] and [AKMMMP]. This description can be viewed as belonging to a more general context of torsors over varieties fibred over the projective line (see the last section of the next chapter).

4

Abelian torsors

This chapter is devoted to the study of various properties of torsors under groups of multiplicative type. The first two sections are technical, and the reader is advised to skip the proofs in the first reading. In Section 4.1 we describe a natural cup-product construction of elements of the Brauer group $\mathrm{Br}(X)$ from X-torsors under groups of multiplicative type. We illustrate this method by constructing the elements of the Brauer group of an elliptic curve. Corollary 4.1.2 will be used in Section 6.1. The language of derived categories is also used in Section 4.2 where we prove the commutativity of a useful diagram of two-term complexes of Γ_k-modules. In Section 4.3, following Colliot-Thélène and Sansuc, we describe the restriction of an X-torsor under a group of multiplicative type S to an appropriate open subset $U \subset X$. This gives a practical recipe for writing down the equations of such a restriction, provided we are given an explicit system of divisors generating $\lambda(\hat{S}) \subset Pic(\overline{X})$, where λ is the type of the torsor. This is demonstrated in Section 4.4 for an important class of torsors such that $\lambda(\hat{S})$ is generated by the components of the 'degenerate' fibres of a dominant morphism $X \to \mathbf{P}^1_k$. This explicit description will be used in Chapter 7 to study the arithmetic of torsors of this type over conic bundle surfaces.

4.1 From abelian torsors to Azumaya algebras

Let X be a k-variety, and S be a k-group of multiplicative type with module of characters $M = \hat{S}$. The following result establishes a connection between sequences (2.22) and (2.23). Recall that $r : \mathrm{Br}_1(X) \to H^1(k, Pic(\overline{X}))$ is the map from the exact sequence (2.23). Let $\lambda : M \to Pic(\overline{X})$ be a homomorphism of Γ_k-modules. Define

$$\mathrm{Br}_\lambda(X) := r^{-1}\lambda_*(H^1(k, M)) \subset \mathrm{Br}_1(X).$$

62

For varieties X satisfying $\overline{k}[X]^* = \overline{k}^*$ we obtain an explicit construction of all elements of $\mathrm{Br}_\lambda(X)$ modulo $\mathrm{Br}_0(X)$ as cup-products via the pairing

$$H^1(k, M) \times H^1(X, S) \xrightarrow{(p^*, id)} H^1(X, M) \times H^1(X, S) \to H^2(X, \mathbf{G}_m) = \mathrm{Br}(X).$$

Theorem 4.1.1 *Let k be a field of characteristic 0 with algebraic closure $\overline{k}$, and $p : X \to Spec(k)$ be a variety over k. Let M be a Γ_k-module finitely generated as an abelian group, S be the dual k-group of multiplicative type, $\lambda \in Hom_k(M, Pic(\overline{X}))$. Suppose that there exists an X-torsor T under S of type λ. Then for $\alpha \in H^1(k, M)$ we have $p^*(\alpha) \cup [T] \in \mathrm{Br}_\lambda(X)$, and*

$$r(p^*(\alpha) \cup [T]) = \lambda_*(\alpha) \quad \in H^1(k, Pic(\overline{X})).$$

If X satisfies $\overline{k}[X]^* = \overline{k}^*$, then any element $A \in \mathrm{Br}_\lambda(X)$ can be represented in the form $A = p^*(\alpha) \cup [T] + p^*(A_0)$ for some $\alpha \in H^1(k, M)$ and $A_0 \in \mathrm{Br}(k)$.

Corollary 4.1.2 *Let X be a variety over the field k of characteristic 0 satisfying $\overline{k}[X]^* = \overline{k}^*$, $\lambda \in Hom_k(M, Pic(\overline{X}))$ for some Γ_k-module M finitely generated as an abelian group. Then for any $A \in \mathrm{Br}_\lambda(X)$ and any $\alpha \in H^1(k, M)$ such that $r(A) = \lambda_*(\alpha)$, there exists $A_0 \in \mathrm{Br}(k)$ such that for any field extension $k \subset K$ and any point $P \in X(K)$ we have*

$$A(P) = Res_{k,K}(\alpha) \cup [T(P)] + Res_{k,K}(A_0) \quad \in \mathrm{Br}(K).$$

This follows from the theorem by the functoriality of the cup-product.

Let us show how Theorem 4.1.1 can be used in practice.

Example: the Brauer group of an elliptic curve.

By a theorem of Grothendieck, $\mathrm{Br}(X)$ injects into the Brauer group of the field of functions $k(X)$ [Grothendieck 68]. The latter is obviously a torsion group, hence the same holds for $\mathrm{Br}(X)$. Thus to describe the elements of $\mathrm{Br}(X)$ it is enough to describe the elements of order n.

Let $X = E$ be an elliptic curve, $M = E[n]$ be the subgroup of points of order n. In Section 3.3 we discussed the E-torsor $n : E \to E$ given by the multiplication by n. Its type was denoted by λ_n; this is the composition of the natural inclusion $E[n](\overline{k}) \hookrightarrow E(\overline{k})$ followed by $E(\overline{k}) = Pic^0(\overline{E}) \hookrightarrow Pic(\overline{E})$. Let $\mathcal{E}_n$ be the class of this torsor in $H^1(E, E[n])$.

Let us compute $\mathrm{Br}(E)_{\lambda_n}$. The degree map gives rise to the exact sequence of Γ_k-modules

$$0 \to E(\overline{k}) \to Pic(\overline{E}) \to \mathbf{Z} \to 0.$$

It is split since $E(k) \neq \emptyset$. We have $H^1(k, \mathbf{Z}) = 0$ implying that $H^1(k, E) = H^1(k, Pic(\overline{E}))$. On the other hand, since E is a curve we have $\mathrm{Br}(\overline{E}) = 0$ (Artin's theorem; see [Grothendieck 68]). Now the exact sequence (2.23) gives us a split exact sequence of abelian groups

$$0 \to \mathrm{Br}(k) \to \mathrm{Br}(E) \to H^1(k, E) \to 0.$$

We identify $H^1(k, E)$ with the subgroup of $\mathrm{Br}(E)$ consisting of elements vanishing at the zero of the group law. In particular, $\mathrm{Br}(E)[n]$ is isomorphic to the direct sum of $\mathrm{Br}(k)[n]$ and $H^1(k, E)[n]$. Finally, the Kummer exact sequence gives rise to the following exact sequence:

$$0 \to E(k)/n \to H^1(k, E[n]) \to H^1(k, E)[n] \to 0.$$

Therefore, $\mathrm{Br}(E)_{\lambda_n}$ is isomorphic to the direct sum of $H^1(k, E)[n]$ and $\mathrm{Br}(k)$, and the projection to $H^1(k, E)[n]$ is precisely the map r.

Theorem 4.1.1 now tells us that from $\alpha \in H^1(k, E[n])$ we build an element

$$A_\alpha := p^*(\alpha) \cup \mathcal{E}_n \ \in \mathrm{Br}(E)_{\lambda_n}.$$

Moreover, the specialization of A_α at the zero of the group law of E is 0, since the corresponding fibre of $n : E \to E$ contains a k-point, hence is a trivial k-torsor. By Theorem 4.1.1 we have $r(A_\alpha) = \lambda_{n*}(\alpha)$, and since $H^1(k, E[n]) \to H^1(k, E)[n]$ is onto, we conclude that *every element of* $\mathrm{Br}(E)_{\lambda_n}$ *or* $\mathrm{Br}(E)[n]$ *vanishing at the zero of the group law equals* A_α *for some* $\alpha \in H^1(k, E[n])$. It is also clear that $A_\alpha = 0$ if and only if α belongs to the image of the Kummer map $E(k)/n \to H^1(k, E[n])$.

When $n = 2$ one can easily find an explicit expression for A_α (see Exercises).

Proof of Theorem 4.1.1. The only non-obvious assertion is the displayed formula. Recall that $Ext_X^n(p^*M, \mathbf{G}_m) = H^n(X, S)$ (Lemma 2.3.7), and also $Ext_X^n(\mathbf{Z}, p^*M) = H^n(X, p^*M)$. There are Yoneda pairings

$$H^1(X, p^*M) \times Ext_X^n(p^*M, \mathbf{G}_m) \to H^{n+1}(X, \mathbf{G}_m) \qquad (4.1)$$

and

$$H^1(k, M) \times Ext_k^n(M, Pic(\overline{X})) \to H^{n+1}(k, Pic(\overline{X})). \qquad (4.2)$$

Let

$$0 \to M \to N \to \mathbf{Z} \to 0 \qquad (4.3)$$

be an exact sequence of Γ_k-modules whose class in $Ext_k^1(\mathbf{Z}, M) = H^1(k, M)$ is α. Let d be the connecting homomorphism in the long exact sequence of Ext's in the first variable associated to (4.3), and also for the pull-back sequence over X. The Yoneda pairings are obtained by splicing extensions, hence for the pairing (4.1) for $\xi \in Ext^n_X(p^*M, \mathbf{G}_m)$ we have $p^*(\alpha) \cup \xi = d(\xi) \in H^{n+1}(X, \mathbf{G}_m)$. In the case of (4.2) we have for $\zeta \in Ext^n_k(M, Pic(\overline{X}))$ the analogous formula $\alpha \cup \zeta = d(\zeta) \in H^{n+1}(k, \mathbf{G}_m)$.

We claim that there is the following commutative diagram:

$$
\begin{array}{ccccc}
H^1(X, S) & = & Ext^1_X(p^*M, \mathbf{G}_m) & \xrightarrow{\;\text{type}\;} & Hom_k(M, Pic(\overline{X})) \\
& & d \downarrow & & d \downarrow \\
& & \mathrm{Br}_1(X) & \xrightarrow{\;\;r\;\;} & H^1(k, Pic(\overline{X}))
\end{array}
\tag{4.4}
$$

Note also that the image of the left vertical map is indeed in $\mathrm{Br}_1(X)$ because (4.3) splits over $\overline{k}$. The right vertical map sends λ to $\lambda_*(\alpha)$. The displayed formula in the lemma is a consequence of the commutativity of this diagram.

In order to prove this commutativity it will be convenient to place ourselves in a more general context. We recall some well known facts and fix our notation (see [Weibel], Chapters 5 and 10, or [GM]). Let $\mathcal{D}^+(X)$, $\mathcal{D}^+(k)$, and $\mathcal{D}^+(Ab)$ be the derived categories of the categories of complexes bounded below: of étale sheaves on X, discrete Γ_k-modules, and abelian groups, respectively, – those categories having enough injectives. If M is an object of a category $\mathcal{C}$, then we denote by $M^\bullet$ the object of $\mathcal{D}^+(\mathcal{C})$ represented by the complex consisting of A in degree zero, and 0 everywhere else. The functor p_* from the category of étale sheaves on X to the category of Γ_k-modules has a left adjoint p^*, $Hom_k(M, p_*F) = Hom_X(p^*M, F)$; this is an exact functor, thus p_* sends injective objects to $Hom_k(M, \cdot)$-acyclic ones. We have a derived functor $\mathbf{R}p_* : \mathcal{D}^+(X) \to \mathcal{D}^+(k)$. The functor $Hom_k(M, \cdot)$ from Γ_k-modules to abelian groups defines a derived functor $\mathbf{R}Hom_k(M, \cdot) : \mathcal{D}^+(k) \to \mathcal{D}^+(Ab)$. If $M = \mathbf{Z}$ we employ the notation $\mathbf{H}(k, \cdot) : \mathcal{D}^+(k) \to \mathcal{D}^+(Ab)$ for the derived functor of $M \mapsto M^{\Gamma_k}$. The composition of $\mathbf{R}p_*$ with $\mathbf{R}Hom_k(M, \cdot)$ is the derived functor $\mathbf{R}Hom_X(p^*M, \cdot)$, and the composition of $\mathbf{R}p_*$ with $\mathbf{H}(k, \cdot)$ is the derived functor $\mathbf{H}(X, \cdot)$. We denote by $\mathbb{H}^i(k, \cdot) = H^i(\mathbf{H}(k, \cdot))$ the hypercohomology functors, and also $\mathbb{R}^i Hom_k(M, \cdot) = H^i \mathbf{R}Hom_k(M, \cdot)$ (this is a hyper-derived functor, see [Milne, EC]; App. C, [Weibel], 5.7).

If F is an object of $\mathcal{D}^+(k)$ and $n \in \mathbf{Z}$, the truncation functors provide the exact triangles $\tau_{\leq n}(F) \to F \to \tau_{\geq n+1}(F)$. We employ the notation $\tau_{[m,n]}(F)$ for $\tau_{\geq m}(\tau_{\leq n}(F))$. Then $\tau_{[n]}(F) := \tau_{[n,n]}(F) = H^n(F)^\bullet[-n]$. An exact sequence of Γ_k-modules

$$
0 \to A \to B \to C \to 0
\tag{4.5}
$$

and an object F of $\mathcal{D}^+(k)$ give rise to the following commutative diagram in $\mathcal{D}^+(Ab)$:

$$\begin{array}{ccc}
\mathbf{R}Hom_k(A, \tau_{\leq 1}(F)) & \longrightarrow & \mathbf{R}Hom_k(A, \tau_{[1]}(F)) \\
d \downarrow & & d \downarrow \\
\mathbf{R}Hom_k(C, \tau_{\leq 1}(F))[1] & \longrightarrow & \mathbf{R}Hom_k(C, \tau_{[1]}(F))[1]
\end{array}$$

Substituting $F = \mathbf{R}p_*\mathbf{G}_m$ into this diagram with the sequence (4.3) as (4.5), and considering the cohomology groups in degree 1, we obtain the commutative diagram of abelian groups:

$$\begin{array}{ccc}
\mathbb{R}^1 Hom_k(M, \tau_{\leq 1}(\mathbf{R}p_*\mathbf{G}_m)) & \longrightarrow & \mathbb{R}^1 Hom_k(M, \tau_{[1]}(\mathbf{R}p_*\mathbf{G}_m)) \\
d \downarrow & & d \downarrow \qquad\qquad (4.6) \\
\mathbb{H}^1(k, \tau_{\leq 1}(\mathbf{R}p_*\mathbf{G}_m)[1]) & \longrightarrow & \mathbb{H}^1(k, \tau_{[1]}(\mathbf{R}p_*\mathbf{G}_m)[1])
\end{array}$$

We claim that this is precisely the desired diagram (4.4).

Let us first identify the corresponding objects of (4.6) and (4.4). The complex $\tau_{\geq 2}(F)$ is acyclic in degrees 0 and 1. Thus it follows from the existence of the exact triangle $\tau_{\leq 1}(F) \to F \to \tau_{\geq 2}(F)$ that

$$\mathbb{R}^1 Hom_k(M, \tau_{\leq 1}(\mathbf{R}p_*\mathbf{G}_m)) = \mathbb{R}^1 Hom_k(M, \mathbf{R}p_*\mathbf{G}_m) = Ext^1_X(p^*M, \mathbf{G}_m).$$

The second equality is due to the fact that the composition of $\mathbf{R}p_*(\cdot)$ with $\mathbf{R}Hom_k(M, \cdot)$ is the functor $\mathbf{R}Hom_X(p^*M, \cdot)$.

Recall that for a complex of Γ_k-modules F there is the hypercohomology spectral sequence $H^p(k, H^q(F)) \Rightarrow \mathbb{H}^{p+q}(k, F)$ ([Milne, EC], App. C, (g), [Weibel], 5.7.10). Applied to $F = \tau_{\geq 2}(\mathbf{R}p_*\mathbf{G}_m)$ it computes

$$\mathbb{H}^2(k, \tau_{\geq 2}(\mathbf{R}p_*\mathbf{G}_m)) = H^2(\overline{X}, \mathbf{G}_m)^{\Gamma_k}.$$

We have $\mathbb{H}^2(k, \mathbf{R}p_*\mathbf{G}_m) = H^2(X, \mathbf{G}_m)$, because $\mathbf{H}(k, \mathbf{R}p_*(F)) = \mathbf{H}(X, F)$. Since $\tau_{\geq 2}(\mathbf{R}p_*\mathbf{G}_m)$ is acyclic in degree 1, we get an exact sequence

$$0 \to \mathbb{H}^2(k, \tau_{\leq 1}(\mathbf{R}p_*\mathbf{G}_m)) \to H^2(X, \mathbf{G}_m) \to H^2(\overline{X}, \mathbf{G}_m)^{\Gamma_k}.$$

This identifies $\mathbb{H}^1(k, \tau_{\leq 1}(\mathbf{R}p_*\mathbf{G}_m)[1]) = \mathbb{H}^2(k, \tau_{\leq 1}(\mathbf{R}p_*\mathbf{G}_m))$ with $\mathrm{Br}_1(X)$. Next, we have $\tau_{[1]}(\mathbf{R}p_*\mathbf{G}_m) = Pic(\overline{X})^\bullet[-1]$, hence

$$\begin{aligned}
\mathbb{R}^1 Hom_k(M, \tau_{[1]}(\mathbf{R}p_*\mathbf{G}_m)) &= H^1(Hom_k(M, Pic(\overline{X}))^\bullet[-1]) \\
&= Hom_k(M, Pic(\overline{X})).
\end{aligned}$$

In a similar way

$$\mathbb{H}^1(k, \tau_{[1]}(\mathbf{R}p_*\mathbf{G}_m)[1]) = \mathbb{H}^1(k, Pic(\overline{X})^\bullet) = H^1(k, Pic(\overline{X})).$$

Let us now show that the arrows in (4.4) and (4.6) are identical. This is clear for the vertical maps. Recall that the upper horizontal map in

(4.4) (the type of the torsor) is the edge map of the Grothendieck spectral sequence of composed functors (2.16). Let us now explain why it is possible to identify the upper horizontal maps of (4.4) and (4.6). The proof for the lower horizontal map is completely analogous. Although this is fairly standard material we give details for the sake of completeness. This will complete the proof of Theorem 4.1.1.

Let $\mathcal{A}$, $\mathcal{B}$, $\mathcal{C}$ be abelian categories such that $\mathcal{A}$ and $\mathcal{B}$ have enough injectives. Let $G : \mathcal{A} \to \mathcal{B}$ and $F : \mathcal{B} \to \mathcal{C}$ be left exact additive functors such that G sends injective objects of $\mathcal{A}$ into F-acyclic objects of $\mathcal{B}$. The Grothendieck spectral sequence of composed functors is the first quadrant spectral sequence (see [Weibel], 5.8)

$$E_2^{pq} = (R^p F)(R^q G)(A) \Rightarrow R^{p+q}(FG)(A). \tag{4.7}$$

Let $Ch^+(\mathcal{A})$ be the category of chain complexes A^* such that $A^p = 0$ for $p \ll 0$. Then there are right hyper-derived functors $\mathbb{R}^i F : Ch^+(\mathcal{A}) \to \mathcal{B}$ ([Weibel], 5.7.9). Finally, there are derived functors $\mathbf{R}F : \mathcal{D}^+(\mathcal{A}) \to \mathcal{D}^+(\mathcal{B})$ and $\mathbf{R}G : \mathcal{D}^+(\mathcal{B}) \to \mathcal{D}^+(\mathcal{C})$ between derived categories of complexes bounded below ([Weibel], 10.5.6). Then $H^i \mathbf{R}F(A) = \mathbb{R}^i F(A)$. We have $\mathbf{R}(FG) = (\mathbf{R}F)(\mathbf{R}G)$ (in the sense that there is a unique natural transformation which is an isomorphism, [Weibel], 10.8), and there is a convergent spectral sequence

$$E_2^{pq} = (R^p F)(\mathbb{R}^q G)(A) \Rightarrow \mathbb{R}^{p+q}(FG)(A), \tag{4.8}$$

where $A \in Ob(Ch^+(\mathcal{A}))$. If A is the complex consisting of one object in degree 0, then (4.8) reduces to (4.7) (*ibidem*).

We claim that if $A \in Ob(Ch^+(\mathcal{A}))$ is such that $A^p = 0$ for $p < 0$ (for instance, one object of $\mathcal{A}$ in degree 0), then the edge map

$$\mathbb{R}^n(FG)(A) \to F(\mathbb{R}^n G)(A) \tag{4.9}$$

can be obtained by applying $\mathbb{R}^n F(\cdot) = H^n \mathbf{R}F(\cdot)$ to the canonical map

$$(\mathbf{R}G)(A) \to \tau_{\geq n}(\mathbf{R}G)(A).$$

Let us recall the construction of (4.8) (cf. [Weibel], 5.8.3). Every chain complex $A^* = (A^i, \partial_i)$ has an injective Cartan–Eilenberg resolution $I = I^{**}$, together with an augmentation $A^* \to I^{0*}$. (It has the property that it induces injective resolutions on $Im(\partial_i)$, $Ker(\partial_i)$, and $H^i A$.) The total complex $Tot(I)$ is quasi-isomorphic to A. By definition ([Weibel], 5.7.9) $\mathbb{R}^i G(A)$ is the i-th cohomology group of the total complex

$Tot(G(I))$ of $G(I)$. By the existence theorem for derived functors ([Weibel], 10.5.6) $Tot(G(I))$ represents $\mathbf{R}G(A)$ in $\mathcal{D}^+(\mathcal{B})$, and $\mathbb{R}^q G(A) = H^q \mathbf{R}G(A)$. Repeating this operation by taking a Cartan–Eilenberg resolution J of $Tot(G(I))$ we get $(\mathbb{R}^i F)(Tot(G(I)))$ as the i-th cohomology group of the total complex of the double complex $F(J)$. There are two spectral sequences attached to a double complex. Recall that by our assumption each $Tot(G(I))^i$ is F-acyclic. The spectral sequence

$$^{II}E_2^{pq} = H_v^p H_h^q F(J) = H^p(\mathbb{R}^q F)(Tot(G(I))) \Rightarrow (\mathbb{R}^{p+q} F)(Tot(G(I)))$$

thus degenerates and yields $(\mathbb{R}^i F)(Tot(G(I))) = \mathbb{R}^i (FG)(A)$. The other spectral sequence

$$^{I}E_2^{pq} = H_h^p H_v^q F(J) = (R^p F)(H^q Tot(G(I))) \Rightarrow (\mathbb{R}^{p+q} F)(Tot(G(I)))$$

becomes the spectral sequence (4.8). The edge map (4.9) is thus the natural map

$$H^n Tot(F(J)) \to {}^{I}E_2^{0n} = H^n Tot({}^{I}\tau_{\geq n}(F(J))) = H^n Tot(F({}^{I}\tau_{\geq n}(J))),$$

where ${}^{I}\tau_{\geq n}$ is the truncation by rows. The right hand side equality is due to the fact that injective objects are F-acyclic, thus $F(J^{i,j}/Im(J^{i,j-1})) = F(J^{i,j})/F(Im(J^{i,j-1}))$. Since ${}^{I}\tau_{\geq n}(J)$ is a Cartan–Eilenberg resolution of $\tau_{\geq n}(Tot(G(I)))$ which represents $\tau_{\geq n}(\mathbf{R}G)(A)$, our claim is proved.

The proof of Theorem 4.1.1 is now complete.   QED

4.2 A commutative diagram

In this section we construct a useful commutative diagram of two-term complexes of Γ_k-modules. The diagrams of abelian groups derived from it by applying the hypercohomology functor will be used in the next section and in Section 6.1.

Let $p : X \to Spec(k)$ be a geometrically integral and smooth k-variety. Let

$$0 \to A \to B \to C \to 0 \tag{4.10}$$

be an exact sequence of étale sheaves on X such that

$$R^1 p_* C = 0.$$

We have an exact triangle $A^\bullet \to B^\bullet \to C^\bullet$ in $\mathcal{D}^+(X)$, the derived category of complexes bounded below of étale sheaves on X. Then we have an exact triangle in $\mathcal{D}^+(k)$, the derived category of discrete Γ_k-modules:

$$\mathbf{R}p_*(A^\bullet) \to \mathbf{R}p_*(B^\bullet) \to \mathbf{R}p_*(C^\bullet).$$

Then we obtain the following diagram in $\mathcal{D}^+(k)$:

$$
\begin{array}{ccccccc}
(p_*A)^\bullet & \to & \tau_{[0,1]}\mathbf{R}p_*(A^\bullet) & \to (R^1p_*A)^\bullet[-1] \to & (p_*A)^\bullet[1] \\
\downarrow & & \downarrow & \downarrow & \downarrow \\
(p_*B)^\bullet & \to & \tau_{[0,1]}\mathbf{R}p_*(B^\bullet) & \to (R^1p_*B)^\bullet[-1] \to & (p_*B)^\bullet[1] \\
\downarrow & & \downarrow & \downarrow & \downarrow \\
(p_*B/p_*A)^\bullet & \to & (p_*C)^\bullet & \to (p_*C/p_*B)^\bullet & \to (p_*B/p_*A)^\bullet[1] \\
\downarrow & & \downarrow & \downarrow \quad * & \downarrow \\
(p_*A)^\bullet[1] & \to & \tau_{[0,1]}\mathbf{R}p_*(A^\bullet)[1] \to & (R^1p_*A)^\bullet & \to & (p_*A)^\bullet[2]
\end{array}
\qquad (4.11)
$$

Note that the middle column is an exact triangle because of our condition $R^1p_*C = 0$. In this diagram all squares are commutative except (*), which is anticommutative [BBD, 1.1.11]. We observe that (4.11) can be represented by a diagram of two-term complexes of Γ_k-modules. Let $\mathcal{A}^\cdot$ (resp. $\mathcal{B}^\cdot$, resp. $\mathcal{C}^\cdot$) be an injective resolution of A (resp. of B, resp. of C). Let $\tau_{[0,1]}$ applied to the complex $K^\cdot = \{K^i, \delta_i\}$, $i \geq 0$, be the two-term complex $K^0 \to Ker(\delta_1)$, and $\tau_{[1]}(K^\cdot)$ be the two-term complex $K^0/Ker(\delta_0) \to Ker(\delta_1)$. Then (4.11) is realized by the following diagram of complexes of Γ_k-modules with exact rows and columns:

$$
\begin{array}{ccccccccc}
 & & 0 & & 0 & & 0 \\
 & & \downarrow & & \downarrow & & \downarrow \\
0 & \to & p_*A & \to & \tau_{[0,1]}p_*(\mathcal{A}^\cdot) & \to & \tau_{[1]}p_*(\mathcal{A}^\cdot) & \to & 0 \\
 & & \downarrow & & \downarrow & & \downarrow \\
0 & \to & p_*B & \to & \tau_{[0,1]}p_*(\mathcal{B}^\cdot) & \to & \tau_{[1]}p_*(\mathcal{B}^\cdot) & \to & 0 \\
 & & \downarrow & & \downarrow & & \downarrow \\
0 & \to & p_*B/p_*A & \to & \tau_{[0,1]}p_*(\mathcal{C}^\cdot) & \to & \mathcal{Y}^\cdot & \to & 0 \\
 & & \downarrow & & \downarrow & & \downarrow \\
 & & 0 & & 0 & & 0
\end{array}
\qquad (4.12)
$$

Here the two-term complex $\mathcal{Y}^\cdot$ is defined to make the diagram commutative. (Again, the middle column is exact because $R^1p_*C = 0$.)

We wish to apply this in the case when (4.10) is the exact sequence

$$1 \to \mathbf{G}_{m,X} \to j_*\mathbf{G}_{m,U} \to \bigoplus_{x \in X^{(1)} \cap (X \setminus U)} i_{x*}\mathbf{Z}_x \to 0, \qquad (4.13)$$

where $j : U \to X$ is the inclusion of U, and $i_x : x \to X$ is the embedding of a point $x \in X$, $x \notin U$, $codim_X(x) = 1$. This sequence exists because X

is smooth, and hence Cartier divisors may be interpreted as Weil Divisors; cf. [Milne, EC], II.3.9. (In a variant (2.20) of (4.13) U is replaced by the generic point of X.) We claim that then (4.11) becomes the following pivotal diagram:

$$
\begin{array}{ccccc}
(\overline{k}[X]^*)^\bullet & \to & \tau_{[0,1]}\mathbf{R}p_*\mathbf{G}_m & \to & Pic(\overline{X})^\bullet[-1] \\
\downarrow & & \downarrow & & \downarrow \\
(\overline{k}[U]^*)^\bullet & \to & \tau_{[0,1]}\mathbf{R}(pj)_*\mathbf{G}_m & \to & Pic(\overline{U})^\bullet[-1] \qquad (4.14) \\
\downarrow & & \downarrow & & \downarrow \\
(\overline{k}[U]^*/\overline{k}[X]^*)^\bullet & \to & Div_{\overline{X}\backslash\overline{U}}(\overline{X})^\bullet & \to & Pic_{\overline{X}\backslash\overline{U}}(\overline{X})^\bullet
\end{array}
$$

Here $Div_{\overline{X}\backslash\overline{U}}(\overline{X})$ denotes the subgroup of $Div(\overline{X})$ generated by the divisors with supports in $\overline{X}\setminus\overline{U}$, and similarly for Pic. To justify our claim we first show that

$$\tau_{[0,1]}\mathbf{R}(pj)_*\mathbf{G}_m = \tau_{[0,1]}\mathbf{R}p_*(j_*\mathbf{G}_m).$$

For this it is enough to show that $p_*j_* = (pj)_*$ which is trivial, and that $R^1p_*(j_*\mathbf{G}_m) = R^1(pj)_*\mathbf{G}_m$, which would follow if we could show that the étale X-sheaf $R^1j_*\mathbf{G}_m = 0$. This is true because X is smooth. (For a point P of X let $\mathcal{O}^{sh}_{X,P}$ be the strict henselization of the local ring of X at P. The map $Pic(\mathcal{O}^{sh}_{X,P}) \to Pic(\mathcal{O}^{sh}_{X,P} \times_X U)$ is surjective since X is smooth. Now the last group is the group of sections of the sheaf $R^1j_*\mathbf{G}_m$ on the étale neighbourhoud $Spec(\mathcal{O}^{sh}_{X,P})$ of P, and the Picard group of a local ring is 0.) The group $R^1(pj)_*\mathbf{G}_m$ is just $Pic(\overline{U})$. By ([Milne, EC], III.2.22) $C = \bigoplus_{x\in X^{(1)}\cap(X\backslash U)} i_{x*}\mathbf{Z}_x$ satisfies the required condition $R^1p_*C = 0$.

Let S be a k-group of multiplicative type with module of characters $\hat{S}$. The functor $Hom_k(\hat{S},\cdot)$ from the category of discrete Γ_k-modules to the category of abelian groups defines a derived functor

$$\mathbf{R}Hom_k(\hat{S},\cdot) : \mathcal{D}^+(k) \to \mathcal{D}^+(Ab).$$

We claim that applying it to (4.14) we obtain the following commutative diagram in $\mathcal{D}^+(Ab)$:

$$
\begin{array}{ccccc}
Ext^i_k(\hat{S},\overline{k}[X]^*) & \to & H^i(X,S) & \to & Ext^{i-1}_k(\hat{S},Pic(\overline{X})) \\
\downarrow & & \downarrow & & \downarrow \\
Ext^i_k(\hat{S},\overline{k}[U]^*) & \to & H^i(U,S) & \to & Ext^{i-1}_k(\hat{S},Pic(\overline{U})) \\
\downarrow & & \downarrow & & \downarrow \\
Ext^i_k(\hat{S},\overline{k}[U]^*/\overline{k}[X]^*) & \to & Ext^i_k(\hat{S},Div_{\overline{X}\backslash\overline{U}}(\overline{X})) & \to & Ext^i_k(\hat{S},Pic_{\overline{X}\backslash\overline{U}}(\overline{X}))
\end{array}
$$

$$(4.15)$$

Here $i = 0$ and 1, and also 2 provided $H^2(X,S)$ is replaced by the kernel of

$H^2(X, S) \to Hom_k(\hat{S}, H^2(\overline{X}, \mathbf{G}_m))$, and similarly for U. The only entries of this diagram which are not immediately obvious are the top and the middle objects in the middle column. We give an argument for X; for U it is just the same. In the previous section we already used the fact that

$$\mathbf{R}Hom_X(p^*\hat{S}, \cdot) = \mathbf{R}Hom_k(\hat{S}, \mathbf{R}p_*(\cdot)).$$

We have $\mathbb{R}^i Hom_X(p^*\hat{S}, \mathbf{G}_m) = Ext^i_X(p^*\hat{S}, \mathbf{G}_m)$ which by virtue of Lemma 2.3.7 is canonically isomorphic to $H^i(X, S)$. Now the exact triangle

$$\mathbf{R}Hom_k(\hat{S}, \tau_{[0,1]}\mathbf{R}p_*\mathbf{G}_m) \to \mathbf{R}Hom_k(\hat{S}, \mathbf{R}p_*\mathbf{G}_m) \to \mathbf{R}Hom_k(\hat{S}, \tau_{\geq 2}\mathbf{R}p_*\mathbf{G}_m)$$

combined with the fact that the last object is acyclic in degrees 0 and 1, and its second cohomology is $Hom_k(\hat{S}, H^2(\overline{X}, \mathbf{G}_m))$, implies what we need.

First application.

Let P be the k-group of multiplicative type dual to $\hat{P} := Pic_{\overline{X}\setminus\overline{U}}(\overline{X})$. Assume that $\overline{k}[X]^* = \overline{k}^*$. We get the following commutative diagram with exact rows and columns:

$$
\begin{array}{ccccc}
 & & & & 0 \\
 & & & & \downarrow \\
 & & Hom_k(\hat{S}, Div_{\overline{X}\setminus\overline{U}}(\overline{X})) & \to & Hom_k(\hat{S}, \hat{P}) \\
 & & \downarrow & & \downarrow \\
0 & \to & Ext^1_k(\hat{S}, \overline{k}^*) & \to & H^1(X, S) \quad \to Hom_k(\hat{S}, Pic(\overline{X})) \\
 & & \downarrow & & \downarrow \\
0 & \to & Ext^1_k(\hat{S}, \overline{k}[U]^*) & \to & H^1(U, S) \quad \to Hom_k(\hat{S}, Pic(\overline{U})) \\
 & & \downarrow & & \downarrow \\
Hom_k(\hat{S}, \hat{P}) & \to Ext^1_k(\hat{S}, \overline{k}[U]^*/\overline{k}^*) & \to & Ext^1_k(\hat{S}, Div_{\overline{X}\setminus\overline{U}}(\overline{X})) &
\end{array}
$$

$$(4.16)$$

This diagram was obtained by Colliot-Thélène and Sansuc [CS87a] by elaborate cohomological computations (without using derived categories). It plays a key rôle in the next section.

Second application.

Now let $\hat{S} = \mathbf{Z}$. Intead of $U \subset X$ we now consider the embedding of the generic point $Spec(k(X))$ into X. Then $\mathrm{Br}_1(k(X)) = H^2(k, \overline{k}(X)^*)$.

Assume that $\overline{k}[X]^* = \overline{k}^*$. Then we get a commutative diagram

$$
\begin{array}{ccccc}
\mathrm{Br}(k) & \to & \mathrm{Br}_1(X) & \to H^1(k, Pic(\overline{X})) \\
\downarrow & & \downarrow & \downarrow \\
H^2(k, \overline{k}(X)^*) & = & H^2(k, \overline{k}(X)^*) \to & 0 \\
\downarrow & & \downarrow & \\
\end{array}
$$
$$
H^1(k, Pic(\overline{X})) \to H^2(k, \overline{k}(X)^*/\overline{k}^*) \to H^2(k, Div(\overline{X}))
$$

$$(4.17)$$

This diagram will be used in Section 6.1.

An analogous diagram in the relative situation when $p : X \to Spec(k)$ is replaced by a surjective morphism to a curve is constructed in [CSS98b].

4.3 Local description of abelian torsors

Let X be a smooth and geometrically integral k-variety such that $\overline{k}[X]^* = \overline{k}^*$. The principal result of this section is the theorem below. It describes how to construct explicitly the restriction of an X-torsor under a group S of multiplicative type to an appropriate open subset of X. It also provides a necessary and sufficient condition for the existence of torsors of a given type λ.

The choice of an open subset $U \subset X$ depends on, though it is not uniquely determined by, $\hat{S}$ and λ. We can always find an open dense subset $U \subset X$ such that the image of λ is contained in the kernel of the restriction map $Pic(\overline{X}) \to Pic(\overline{U})$. For this purpose it suffices to find a finite Γ_k-invariant collection of effective divisors on $\overline{X}$ such that $\lambda(\hat{S})$ is contained in the Γ_k-submodule of $Pic(\overline{X})$ generated by them, and then define U as the complement to these divisors in X. This is certainly possible since $\hat{S}$ is a finitely generated abelian group. Let P be the k-group of multiplicative type dual to the Γ_k-module $\hat{P} = Ker[Pic(\overline{X}) \to Pic(\overline{U})]$, so that there is an exact sequence of Γ_k-modules (the third arrow is surjective because X is smooth):

$$0 \to \hat{P} \to Pic(\overline{X}) \to Pic(\overline{U}) \to 0. \qquad (4.18)$$

Consider also another exact sequence of Γ_k-modules:

$$1 \to \overline{k}[U]^*/\overline{k}^* \xrightarrow{\mathrm{div}} Div_{\overline{X}\setminus\overline{U}}(\overline{X}) \to \hat{P} \to 0. \qquad (4.19)$$

Note in passing that $Div_{\overline{X}\setminus\overline{U}}(\overline{X})$ has a Γ_k-invariant base, that is, it is a permutation Γ_k-module.

We claim that there exists an exact sequence of k-groups of multiplicative type

$$1 \to S \to M \to R \to 1 \qquad (4.20)$$

with the properties

(1) M is a quasi-trivial k-torus (by definition this means that $\hat{M}$ has a Γ_k-invariant basis), and

(2) the dual sequence of Γ_k-modules fits into a commutative diagram

$$
\begin{array}{ccccccccc}
0 & \to & \hat{R} & \to & \hat{M} & \to & \hat{S} & \to & 0 \\
 & & \rho \downarrow & & \downarrow & & \lambda \downarrow & & \\
1 & \to & \overline{k}[U]^*/\overline{k}^* & \to & Div_{\overline{X} \backslash \overline{U}}(\overline{X}) & \to & \hat{P} & \to & 0
\end{array}
\qquad (4.21)
$$

To construct the upper row of (4.21) note that for any Γ_k-module N of finite type there exists a permutation Γ_k-module N' with a surjective map $N' \to N$. Applying this to N defined as the fibre product of $Div_{\overline{X} \backslash \overline{U}}(\overline{X})$ and $\hat{S}$ over $\hat{P}$ we set $\hat{M} = N'$ and define the map $\hat{M} \to \hat{S}$ as the composition of the surjective map $\hat{M} \to N$ with the second projection $N \to \hat{S}$. Then ρ is defined as the map such that (4.21) commutes. (Note that in practice the construction of $\hat{M}$ can become rather involved; this is the main complication of this method.)

We can consider M as an R-torsor under S. Let us call it $\mathcal{M}_R$.

Any morphism $\phi : U \to R$ of a k-variety $q : U \to Spec(k)$ to a k-group of multiplicative type R defines a map of Γ_k-modules $\hat{\phi} : \hat{R} \to \overline{k}[U]^*$. In fact, we have

$$
\begin{aligned}
& Hom_k(\hat{R}, \overline{k}[U]^*) = Hom_{U_{\acute{e}t}}(q^*\hat{R}, \mathbf{G}_{m,U}) \\
= \; & H^0(U_{\acute{e}t}, \mathcal{H}om_{U\text{-groups}}(q^*\hat{R}, \mathbf{G}_{m,U})) = H^0(U_{\acute{e}t}, q^*R) = R(U).
\end{aligned}
$$

Here the first equality is the canonical isomorphism due to the fact that q^* is the left adjoint to q_*. By the duality of commutative U-groups $q^*\hat{R}$ and q^*R we have

$$\mathcal{H}om_{U\text{-groups}}(q^*\hat{R}, \mathbf{G}_{m,U}) = q^*R.$$

It is a local verification that this Hom coincides with the Hom of sheaves of abelian groups on $U_{\acute{e}t}$ as in the second term of the displayed formula.

Thus the maps $\phi : U \to R$ bijectively correspond to Γ_k-homomorphisms $\hat{\phi} : \hat{R} \to \overline{k}[U]^*$.

To each map $\phi : U \to R$ we associate a U-torsor $\phi^*(\mathcal{M}_R)$ under S.

Theorem 4.3.1 (Colliot-Thélène–Sansuc) *Let X be a geometrically* *integral smooth k-variety such that $\overline{k}[X]^* = \overline{k}^*$. Let S be a k-group of multiplicative type, and let $\lambda \in Hom_k(\hat{S}, Pic(\overline{X}))$. There exist a dense open set $j : U \subset X$ such that the composition $\hat{S} \to Pic(\overline{X}) \to Pic(\overline{U})$ is 0, and an exact sequence (4.20) subject to conditions (1) and (2) above. Suppose that such an open set U and a sequence (4.20) are fixed. If Y is an X-torsor under S of type λ, then $j^*(Y) = \phi^*(\mathcal{M}_R)$ for a morphism $\phi : U \to R$ such that $\hat{\phi} : \hat{R} \to \overline{k}[U]^*$ is a lifting of $-\rho$, where ρ is defined in (4.21). This correspondence between the set of isomorphism classes of X-torsors under S of type λ and the set of liftings of ρ is surjective. It is bijective if and only if the natural map*

$$Ext_k^1(\hat{S}, \overline{k}^*) \to Ext_k^1(\hat{S}, \overline{k}[U]^*)$$

is injective. This is the case when the following exact sequence of Γ_k-modules is split:

$$1 \to \overline{k}^* \to \overline{k}[U]^* \to \overline{k}[U]^*/\overline{k}^* \to 1. \tag{4.22}$$

This formulation is rather involved but useful in practice. Some of the applications of this theorem can be found in the next section.

The proof of the theorem will occupy the rest of this section.

Here is the general scheme of the proof. We already explained how to construct U and (4.20) with the required properties. Let us consider the following diagram:

$$\begin{array}{ccc}
Hom_k(\hat{R}, \overline{k}[U]^*) & = & H^0(U_{\acute{e}t}, R) \\
\partial \downarrow & & \partial' \downarrow \\
Ext_k^1(\hat{S}, \overline{k}[U]^*) & \xrightarrow{\ \epsilon\ } & H^1(U_{\acute{e}t}, S)
\end{array} \tag{4.23}$$

where ∂ is the differential in the long exact sequence of Ext's associated to the upper row of (4.21), ∂' is the differential in the long exact sequence of cohomology associated to (4.20), and ϵ is the second arrow in (2.21) (the canonical injective map from the spectral sequence (2.16)). We claim that (4.23) commutes. The proof will be given later. As we noticed in Section 3.2 (exact sequence (3.2)), the class of the R-torsor $\mathcal{M}_R$ in $H^1(R_{\acute{e}t}, S)$ is $\partial'(Id)$, where $Id \in H^0(S_{\acute{e}t}, S)$ is the identity map. Hence the class of the U-torsor $\phi^*(\mathcal{M}_R)$ in $H^1(U_{\acute{e}t}, S)$ is $\partial'(\phi)$. The commutativity of (4.23) implies that

$$[\phi^*(\mathcal{M}_R)] = \epsilon(\partial(\hat{\phi})). \tag{4.24}$$

Consider the following obvious commutative diagram with exact rows and columns:

$$0 = Ext^1_k(\hat{M}, \overline{k}^*) \leftarrow Hom_k(\hat{M}, \overline{k}[U]^*/\overline{k}^*) \xleftarrow{\pi} Hom_k(\hat{M}, \overline{k}[U]^*)$$

$$Hom_k(\hat{R}, \overline{k}[U]^*/\overline{k}^*) \xleftarrow{\pi} Hom_k(\hat{R}, \overline{k}[U]^*)$$

$$\partial \downarrow \qquad\qquad\qquad \partial \downarrow$$

$$Ext^1_k(\hat{S}, \overline{k}[U]^*/\overline{k}^*) \xleftarrow{\pi} Ext^1_k(\hat{S}, \overline{k}[U]^*)$$

$$Ext^1_k(\hat{M}, \overline{k}[U]^*/\overline{k}^*) \xleftarrow{\pi} Ext^1_k(\hat{M}, \overline{k}[U]^*) \leftarrow Ext^1_k(\hat{M}, \overline{k}^*) = 0$$

$$(4.25)$$

Here the columns come from the dual sequence of (4.20), and the rows come from (4.22); in particular, the maps π are induced by the natural surjection $\overline{k}[U]^* \to \overline{k}[U]^*/\overline{k}^*$. We have $Ext^1_k(\hat{M}, \overline{k}^*) = H^1(k, M) = 0$ by Hilbert's Theorem 90 combined with Shapiro's lemma, since by construction $\hat{M}$ is isomorphic to a direct sum of modules $\mathbf{Z}[\Gamma_k/\Gamma_K]$, where K is a finite extension of k. Note that $\hat{\phi} \in Hom_k(\hat{R}, \overline{k}[U]^*)$ is a lifting of ρ precisely when $\pi(\hat{\phi}) = \rho$. It follows from the commutativity of (4.25) that the set of elements of $Ext^1_k(\hat{S}, \overline{k}[U]^*)$ of the form $\partial(\hat{\phi})$, where $\hat{\phi}$ is a lifting of ρ, is a subset of $\pi^{-1}(\partial(\rho))$. Since the bottom π is injective and the upper π is surjective, an easy diagram chase shows that these sets actually coincide. Thus we obtain that the set of classes $[\phi^*(\mathcal{M}_R)] \in H^1(U_{\acute{e}t}, S)$ such that $\hat{\phi}$ is a lifting of ρ, is the set $\epsilon(\pi^{-1}(\partial(\rho)))$.

Now note that $\partial(\rho)$ is the push-forward of the upper row of (4.21), and it follows from the commutativity of (4.21) that the same extension of $\hat{S}$ by $\overline{k}[U]^*/\overline{k}^*$ is obtained as the pull-back of (4.19) by λ. We write this as $\partial(\rho) = \delta(\lambda)$, where δ is the differential in the long exact sequence of Ext's in the second variable asociated to (4.19).

Thus we can summarize the preceding discussion as follows: the set of classes $[\phi^*(\mathcal{M}_R)] \in H^1(U_{\acute{e}t}, S)$ such that $\hat{\phi}$ is a lifting of ρ is the set $\epsilon(\pi^{-1}(\delta(\lambda)))$.

By the results of the previous section we have the following commutative diagram with exact rows and columns:

$$
\begin{array}{ccccc}
& & & & 0 \\
& & & & \downarrow \\
& & Hom_k(\hat{S}, Div_{\overline{X}\setminus \overline{U}}(\overline{X})) & \longrightarrow & Hom_k(\hat{S}, \hat{P}) \\
& & \downarrow & & \downarrow i \\
0 & \longrightarrow & Ext_k^1(\hat{S}, \overline{k}^*) & \longrightarrow & H^1(X_{\acute{e}t}, S) \xrightarrow{\text{type}} Hom_k(\hat{S}, Pic(\overline{X})) \\
& & \downarrow & & \downarrow j^* \\
0 & \longrightarrow & Ext_k^1(\hat{S}, \overline{k}[U]^*) & \xrightarrow{\epsilon} & H^1(U_{\acute{e}t}, S) \xrightarrow{\text{type}} Hom_k(\hat{S}, Pic(\overline{U})) \\
& & \downarrow \pi & & \downarrow \\
Hom_k(\hat{S}, \hat{P}) & \xrightarrow{\delta} & Ext_k^1(\hat{S}, \overline{k}[U]^*/\overline{k}^*) & \longrightarrow & Ext_k^1(\hat{S}, Div_{\overline{X}\setminus \overline{U}}(\overline{X}))
\end{array}
$$

$$(4.26)$$

The middle rows here are sequences (2.22) and (2.21); the upper and the bottom rows are pieces of the long exact sequence of Ext's associated to (4.19). The left hand (resp. the right hand) column is a piece of the long exact sequence of Ext's associated to (4.22) (resp. to (4.18)).

Let us think of λ as an element of $Hom_k(\hat{S}, \hat{P})$; this should not lead to confusion since the tautological map $i : Hom_k(\hat{S}, \hat{P}) \to Hom_k(\hat{S}, Pic(\overline{X}))$ is injective. Suppose we have an X-torsor Y under S of type λ, in other words, a class $[Y] \in H^1(X_{\acute{e}t}, S)$ such that $\text{type}(Y) = i(\lambda)$. Then the restriction $[j^*(Y)] \in H^1(U_{\acute{e}t}, S)$ comes from an element $\zeta \in Ext_k^1(\hat{S}, \overline{k}[U]^*)$, $\epsilon(\zeta) = [j^*(Y)]$, as is clear from the commutativity of (4.26).

We claim that

$$\pi(\zeta) = -\delta(\lambda).$$

This implies that $\zeta \in \pi^{-1}(\delta(-\lambda)) = \pi^{-1}(\partial(-\rho))$, that is, $j^*(Y) = \phi^*(\mathcal{M}_R)$ for some $\phi : U \to R$ such that $\hat{\phi}$ is a lifting of $-\rho$.

Let us prove that any U-torsor $\phi^*(\mathcal{M}_R)$ where $\phi : U \to R$ is such that $\hat{\phi}$ is a lifting of $-\rho$ extends to an X-torsor of type λ. The fact that it does extend immediately follows from commutativity of (4.26), since the image of $\partial(\rho) = \delta(\lambda) \in Ext_k^1(\hat{S}, \overline{k}[U]^*/\overline{k}^*)$ in $Ext_k^1(\hat{S}, Div_{\overline{X}\setminus \overline{U}}(\overline{X}))$ is 0. Suppose that the X-torsor Y' so obtained has type $\lambda' \in Hom_k(\hat{S}, \hat{P})$. The analogue of the last displayed formula tells us that then $\delta(\lambda - \lambda') = 0$, hence $\lambda - \lambda'$ comes from an element of $Hom_k(\hat{S}, Div_{\overline{X}\setminus \overline{U}}(\overline{X}))$. Modifying Y' by the image of this element in $H^1(X_{\acute{e}t}, S)$ we see that our U-torsor can indeed be extended to an X-torsor of type λ. This proves the surjectivity statement in the theorem. The bijectivity statement is obvious from the commutativity of (4.26).

It remains to prove that diagram (4.23) commutes, and to show that $\pi(\zeta) = -\delta(\lambda)$.

Commutativity of (4.23). The basic underlying observation is that the functor

$$\mathbf{R}\mathit{Hom}_X(p^*\hat{R}, \cdot) : \mathcal{D}^+(X) \to \mathcal{D}^+(Ab)$$

can be represented as a composition of functors in two different ways:

$$\mathbf{R}\mathit{Hom}_X(p^*\hat{R}, \cdot) = \mathbf{H}(X, \mathbf{R}\mathcal{H}om_X(p^*\hat{R}, \cdot)),$$

and

$$\mathbf{R}\mathit{Hom}_X(p^*\hat{R}, \cdot) = \mathbf{R}\mathit{Hom}_k(\hat{R}, \mathbf{R}p_*(\cdot)).$$

Both representations give rise to the Grothendieck spectral sequences of composed functors; the first one was used in the proof of Lemma 2.3.7, and the second one is (2.16). We have $\mathcal{H}om_X(p^*\hat{R}, \mathbf{G}_m) = p^*R$. By Sublemma 2.3.8 we have $\mathcal{E}xt^i_X(p^*\hat{R}, \mathbf{G}_m) = 0$ for $i > 0$, hence the first spectral sequence completely degenerates. The edge $E^2_{n,0} \to E_n$ of any spectral sequence of composed functors $\mathbf{R}\Psi \circ \mathbf{R}\Phi = \mathbf{R}\Psi\Phi$ can be obtained by passing to H^n in the natural map

$$\mathbf{R}\Psi(\Phi(\cdot)^\bullet) = \mathbf{R}\Psi(\tau_{\leq 0}\mathbf{R}\Phi(\cdot)) \to \mathbf{R}\Psi(\mathbf{R}\Phi(\cdot)) = \mathbf{R}\Psi\Phi(\cdot).$$

The upper row of (4.21) gives rise to compatible natural maps

$$\mathbf{R}\mathit{Hom}_k(\hat{R}, \cdot) \to \mathbf{R}\mathit{Hom}_k(\hat{S}, \cdot)[1], \quad \mathbf{R}\mathcal{H}om_X(p^*\hat{R}, \cdot) \to \mathbf{R}\mathcal{H}om_X(p^*\hat{S}, \cdot)[1].$$

Their functoriality implies the commutativity of the following diagram:

$$
\begin{array}{ccccc}
\mathbf{R}\mathit{Hom}_k(\hat{R}, \tau_{\leq 0}\mathbf{R}p_*\mathbf{G}_m) & \to & \mathbf{R}\mathit{Hom}_X(p^*\hat{R}, \mathbf{G}_m) & \leftarrow & \mathbf{H}(X, \mathcal{H}om_X(p^*\hat{R}, \mathbf{G}_m)) \\
\downarrow & & \downarrow & & \downarrow \\
\mathbf{R}\mathit{Hom}_k(\hat{S}, \tau_{\leq 0}\mathbf{R}p_*\mathbf{G}_m)[1] & \to & \mathbf{R}\mathit{Hom}_X(p^*\hat{S}, \mathbf{G}_m)[1] & \leftarrow & \mathbf{H}(X, \mathcal{H}om_X(p^*\hat{S}, \mathbf{G}_m))[1]
\end{array}
$$

Passing to H^0 we obtain a commutative diagram where the vertical arrows are the differentials associated to the upper row of (4.21):

$$
\begin{array}{ccccc}
\mathit{Hom}_k(\hat{R}, p_*\mathbf{G}_m) & \to & \mathit{Hom}_X(p^*\hat{R}, \mathbf{G}_m) & \leftarrow & H^0(X, R) \\
\downarrow & & \downarrow & & \downarrow \\
\mathit{Ext}^1_k(\hat{S}, p_*\mathbf{G}_m) & \to & \mathit{Ext}^1_X(p^*\hat{S}, \mathbf{G}_m) & \leftarrow & H^1(X, S)
\end{array}
$$

Since the first of our two spectral sequences degenerates, the right horizontal arrows are isomorphisms. This proves our claim.

The proof of $\pi(\zeta) = -\delta(\lambda)$. This is a particular case of the following statement. Let

$$
\begin{array}{ccccccccc}
 & & 0 & & 0 & & 0 & & \\
 & & \downarrow & & \downarrow & & \downarrow & & \\
0 & \to & A_{11} & \to & A_{12} & \to & A_{13} & \to & 0 \\
 & & \downarrow & & \downarrow & & \downarrow & & \\
0 & \to & A_{21} & \to & A_{22} & \to & A_{23} & \to & 0 \\
 & & \downarrow & & \downarrow & & \downarrow & & \\
0 & \to & A_{31} & \to & A_{32} & \to & A_{33} & \to & 0 \\
 & & \downarrow & & \downarrow & & \downarrow & & \\
 & & 0 & & 0 & & 0 & &
\end{array}
\tag{4.27}
$$

be a commutative diagram of complexes in an abelian category with exact rows and columns. We denote the horizontal arrows in this diagram by h, and the vertical ones by v. The differentials in the long exact sequences of cohomology asociated to rows (resp. to columns) will be denoted by ∂_h (resp. by ∂_v).

Lemma 4.3.2 *Let $i \geq 1$. Suppose we are given $\beta \in H^i(A_{21})$ and $\gamma \in H^i(A_{12})$ such that $h_*(\beta) = v_*(\gamma)$. Then there exists $\alpha \in H^{i-1}(A_{33})$ such that $v_*(\beta) = \partial_h(\alpha)$ and $h_*(\gamma) = -\partial_v(\alpha)$.*

To deduce the required formula let $i = 1$. Note that (4.14) is obtained from a commutative diagram of complexes (4.12). Hence we let (4.27) be the diagram of complexes obtained from (4.14) by applying the derived functor $\mathbf{R}Hom_k(\hat{S}, \cdot)$. In the notation of (4.26) let $\gamma = [Y]$ be such that $\mathtt{type}(Y) = i(\lambda)$, and let $\beta = \zeta$. The assumption of the lemma is satisfied because $\epsilon(\zeta) = [j^*(Y)]$. In our case $\partial_v = i$ is injective, thus $\alpha \in Hom_k(\hat{S}, \hat{P})$ provided by the lemma equals λ. Now the last formula of the lemma proves what we need.

Proof of Lemma 4.3.2. Define A as the quotient of $A_{21} + A_{12}$ by the image of A_{11} embedded by $x \mapsto (v(x), -h(x))$. Then we have a commutative diagram with exact rows:

$$
\begin{array}{ccccccccc}
0 & \to & A_{13} & \to & A_{23} & \to & A_{33} & \to & 0 \\
 & & \uparrow & & \uparrow & & \| & & \\
0 & \to & A & \to & A_{22} & \to & A_{33} & \to & 0 \\
 & & \downarrow & & \downarrow & & \| & & \\
0 & \to & A_{31} & \to & A_{32} & \to & A_{33} & \to & 0
\end{array}
$$

where $A \to A_{13}$ is induced by the natural map $A_{21} + A_{12} \to A_{12} \to A_{13}$, and similarly for $A \to A_{31}$. The map $A \to A_{22}$ is induced by the sum of

$A_{21} \to A_{22}$ and $A_{12} \to A_{22}$. Let $\xi \in H^i(A)$ be the image of $(\beta, -\gamma) \in$ $H^i(A_{21} + A_{12})$. By the assumption ξ goes to 0 in $H^i(A_{22})$, hence $\xi = \partial(\alpha)$ for some $\alpha \in H^{i-1}(A_{33})$. Observe that the image of ξ in $H^i(A_{13})$ is $-h(\gamma)$, and the image of ξ in $H^i(A_{31})$ is $v(\beta)$. Now the conclusion of the lemma follows from the commutativity of the last diagram.   QED

The proof of the theorem is now complete.

Remark. Let us point out an important particular case when the vertical arrows in (4.21) are identity maps. Then the theorem says that X-torsors of such a type bijectively correspond to Γ_k-splittings of (4.22). An example of this situation is universal X-torsors, characterized by the property that $Pic(\overline{U}) = 0$.

Since for varieties X such that $Pic(\overline{X})$ is of finite type, the universal torsors are the most natural torsors one could think of, it would be desirable to be able to construct them explicitly. The main difficulty in using the local description of universal torsors is due to the fact that in general there is no 'canonical' Γ_k-invariant system of generators of $Pic(\overline{X})$, or, equivalently, no 'canonical' dense open subset U such that $Pic(\overline{U}) = 0$. The notable exceptions are toric varieties, or, more generally, smooth k-varieties X such that $\overline{k}[X]^* = \overline{k}^*$ and X contains a k-torsor U under an algebraic k-torus (see Section 6.3). Note also that the geometric Picard group of Del Pezzo surfaces of degree less than 8 is generated by the classes of exceptional curves. (The example of Del Pezzo surfaces of degree 5 was treated in the appendix to Section 3.1). Explicit equations of universal torsors are known for diagonal cubic surfaces [CKS]. For conic bundles over $\mathbf{P}_k^1$ one can explicitly describe torsors which are closely related to universal torsors. This is a particular case of a more general set-up of the next section.

4.4 Torsors associated with a dominant morphism to $\mathbf{P}_k^1$

We now describe a useful class of torsors that can be defined whenever a geometrically integral smooth k-variety X is equipped with a surjective morphism f onto $\mathbf{P}_k^1$. (Then f is faithfully flat.) We keep the·assumption $\overline{k}[X]^* = \overline{k}^*$.

Let $U' \subset \mathbf{P}_k^1$ be a dense open set. We define $U = f^{-1}(U')$. Let S be the k-group of multiplicative type such that

$$\hat{S} = Div_{\overline{X}\setminus\overline{U}}(\overline{X})/(\overline{k}[U']^*/\overline{k}^*).$$

Note that although we take all the divisors supported outside $\overline{U}$, we allow only the relations given by the rational functions coming from invertible functions on $U' \subset \mathbf{P}^1_k$. We denote by $\lambda : \hat{S} \to Pic(\overline{X})$ the map sending a divisor to its class. The kernel of λ is $\overline{k}[U]^*/\overline{k}[U']^*$, which is trivial if the generic fibre of f has no other invertible functions than those coming from the rational functions on $\mathbf{P}^1_k$, for example, if the generic fibre is a proper and geometrically integral variety. This kernel can be non-trivial when f is finite. Although we started with an arbitrary dense open set $U' \subset \mathbf{P}^1_k$, for meaningful applications we shall later choose U' with the property that the fibres of f over U' are 'non-degenerate' (e.g., the restriction of f to U' is smooth, or some weaker condition).

The aim of this section is to describe by explicit equations the restriction of X-torsors under S of type λ to the open subset of U obtained by removing one more fibre of f. We begin by showing that the existence of torsors of this type imposes no condition on X.

Proposition 4.4.1 *For any X, S, λ as above, the X-torsors under S of type λ exist.*

Proof. The following obvious diagram of Γ_k-modules commutes:

$$
\begin{array}{ccccccccc}
1 & \to & \overline{k}^* & \to & \overline{k}[U']^* & \to & Div_{\overline{X}\setminus\overline{U}}(\overline{X}) & \to & \hat{S} & \to & 0 \\
 & & \| & & \downarrow & & \downarrow & & \lambda \downarrow & & \\
0 & \to & \overline{k}^* & \to & \overline{k}(X)^* & \to & Div(\overline{X}) & \to & Pic(\overline{X}) & \to & 0
\end{array}
$$
$$(4.28)$$

Theorem 2.3.4 *(a)* and the commutativity of (4.28) imply that $\lambda^*(e(X)) \in Ext^2_k(\hat{S}, \overline{k}^*)$ is represented by the upper extension (up to sign). But U' is an open subset of the proper variety $\mathbf{P}^1_k$ with k-points, hence by Theorem 2.3.4 *(b)* the injection $\overline{k}^* \to \overline{k}[U']^*$ has a Γ_k-equivariant section. Thus $\lambda^*(e(X)) = 0$, and by the results of Section 2.3 we get $\partial(\lambda) = 0$ which means that X-torsors of type λ exist. QED

We need some more notation. We choose a k-point in U', which is always possible if k is infinite. We denote it by ∞, and set $\mathbf{A}^1_k = \mathbf{P}^1_k \setminus \{\infty\}$ with coordinate x, $X_\infty = f^{-1}(\infty)$, $U'_0 = U' \setminus \{\infty\}$, $U_0 = f^{-1}(U'_0) = U \setminus X_\infty$.

Let $T = \mathbf{P}^1_k \setminus U'$ be the complement of U'. We shall denote by Σ the set of irreducible components s of the fibres of f at the closed points $t \in T$. We use the notation $k_t = k(t)$ for the residue field at $t \in T$, $k(s)$ for the field of functions of the component $s \in \Sigma$, and k_s for the integral closure of k_t in $k(s)$ where $t = f(s)$.

Let $P_t(x)$ be the monic irreducible polynomial defining t; then $k_t = $

$k[a_t] = k[x]/(P_t(x))$ (by a_t we denote the image of x in k_t). Let $d_t = [k_t : k]$, and let

$$P_t(x) = \prod_{i=1}^{d_t}(x - e_{t,i}),$$

where $e_{t,i} \in \overline{k}$. Let $\psi_{t,i}$, for $i = 1, \ldots, d_t$, be the embedding of k_t into $\overline{k}$ which sends a_t to $e_{t,i}$. By E we shall denote the set of points in $\mathbf{P}_{\overline{k}}^1$ with coordinates $e_{t,i}$, and by C the set of irreducible components of the corresponding $\overline{k}$-fibres of f. We shall sometimes denote by $x(e)$ the coordinate of e.

For $e \in E$ we have

$$div(x - x(e)) = X_\infty - \sum_{f(c)=e} m(c)c \; \in Div(\overline{X}), \qquad (4.29)$$

where $m(c)$ is the multiplicity of c in its fibre. By this we mean that up to functions invertible in the local ring of c, a uniformizing parameter at e is the $m(c)$-th power of a local equation of c. Since the multiplicities of the components interchanged by Γ_k are clearly the same, we shall write $m(s)$ for $m(c)$, where c is any of the components into which s splits over $\overline{k}$.

Lemma 4.4.2 *The abelian group $\hat{S}$ is generated by the elements γ_c, for $c \in C$, and γ_∞, such that all the relations between them are linear combinations of $\gamma_\infty - \sum_{f(c)=e} m(c)\gamma_c = 0$, $e \in E$.*

Proof. This is clear from (4.29) and the definition of $\hat{S}$. QED

This description makes it possible to choose (4.21) so that

$$\hat{R} = \overline{k}[U_0']^*/\overline{k}^* = \bigoplus_{e \in E} \mathbf{Z}, \quad \text{and} \quad \hat{M} = \mathbf{Z}X_\infty \oplus Div_{\overline{X}\setminus\overline{U}}(\overline{X}) = \mathbf{Z} \oplus \bigoplus_{c \in C} \mathbf{Z}$$

with the natural permutation action of the Galois group Γ_k. The homomorphism $div : \hat{R} \to \hat{M}$ sends the class of the function $x - x(e)$ to the divisor $-X_\infty + \sum_{f(c)=e} m(c)c$. In terms of Γ_k-invariant bases, div sends 1_e to $-1 \oplus \sum_{f(c)=e} m(c)1_c$. The homomorphism $\hat{M} \to \hat{S}$ sends the generator of the first $\mathbf{Z}$ to γ_∞, and 1_c to γ_c. Thus (4.21) takes the form

$$
\begin{array}{ccccccccc}
0 & \to & \bigoplus_{e \in E} \mathbf{Z} & \xrightarrow{div} & \mathbf{Z} \oplus \bigoplus_{c \in C} \mathbf{Z} & \to & \hat{S} & \to & 0 \\
 & & \| & & \| & & \| & & \\
0 & \to & \overline{k}[U_0']^*/\overline{k}^* & \xrightarrow{div} & \mathbf{Z}X_\infty \oplus Div_{\overline{X}\setminus\overline{U}}(\overline{X}) & \to & \hat{S} & \to & 0 \quad (4.30) \\
 & & f^* \downarrow & & \downarrow & & \lambda \downarrow & & \\
1 & \to & \overline{k}[U_0]^*/\overline{k}^* & \xrightarrow{div} & Div_{\overline{X}\setminus\overline{U_0}}(\overline{X}) & \to & \hat{P} & \to & 0
\end{array}
$$

Recall that $\hat{P}$ is the kernel of the restriction $Pic(\overline{X}) \to Pic(\overline{U_0})$. The left vertical map f^* in (4.30) is the natural injection $\overline{k}[U_0']^*/\overline{k}^* \to \overline{k}[U_0]^*/\overline{k}^*$, the middle vertical map is the natural injection. Note that $Ker(\lambda) = \overline{k}[U_0]^*/\overline{k}[U_0']^*$ need not be trivial in general. It is trivial, however, if all the invertible functions on the generic fibre of f come from $\overline{k}(\mathbf{P}_k^1)$ (for instance, if the generic fibre is proper and geometrically integral).

The sequence of $\overline{k}$-groups of multiplicative type dual to the upper row of (4.30) is

$$1 \to \overline{S} \to \mathbf{G}_{m,\overline{k}} \times \prod_{c \in C} \mathbf{G}_{m,\overline{k}} \to \prod_{e \in E} \mathbf{G}_{m,\overline{k}} \to 1, \qquad (4.31)$$

where the surjective map sends $(x; x_c, c \in C)$ to $(x^{-1} \prod_{f(c)=e} x_c^{m(c)}, e \in E)$.

To a finite field extension K/k one associates a k-torus obtained by the Weil descent of scalars from $\mathbf{G}_{m,K}$, it is denoted by $R_{K/k}(\mathbf{G}_{m,K})$. This torus can be defined by the property that for any k-algebra L one has $R_{K/k}(\mathbf{G}_{m,K})(L) = L \otimes_k K$. The set of $\overline{k}$-points of $R_{K/k}(\mathbf{G}_{m,K})$ is

$$(\overline{k} \otimes_k K)^* \cong (\overline{k}^*)^n,$$

where $n = [K : k]$, and the isomorphism $\overline{k} \otimes_k K \cong \overline{k}^n$ sends $x \otimes y$ to $(x\psi_1(y), \dots, x\psi_n(y))$, where $\psi_1, \dots, \psi_n$ are the different embeddings of K into $\overline{k}$. If $k \subset K \subset K'$ are finite extensions, there are two natural maps:

$$N_{K'/K} : R_{K'/k}(\mathbf{G}_{m,K'}) \longrightarrow R_{K/k}(\mathbf{G}_{m,K}),$$

given on the sets of $\overline{k}$-points $(\overline{k} \otimes_k K')^* \to (\overline{k} \otimes_k K)^*$ by the norm from K' to K, and

$$i_{K'/K} : R_{K/k}(\mathbf{G}_{m,K}) \longrightarrow R_{K'/k}(\mathbf{G}_{m,K'}),$$

given on the sets of $\overline{k}$-points $(\overline{k} \otimes_k K)^* \to (\overline{k} \otimes_k K')^*$ by the inclusion $K \hookrightarrow K'$.

In these terms the sequence (4.20) of the previous section can be written as

$$1 \to S \to \mathbf{G}_{m,k} \times \prod_{s \in S} R_{k_s/k}(\mathbf{G}_{m,k_s}) \to \prod_{t \in T} R_{k_t/k}(\mathbf{G}_{m,k_t}) \to 1. \qquad (4.32)$$

Here the restriction of the surjective map to $\mathbf{G}_{m,k}$ is $x \mapsto x^{-1}$ followed by $(i_{k_t/k}, t \in T)$, and the restriction of this map to $R_{k_s/k}(\mathbf{G}_{m,k_s})$ is $N_{k_s/k_t}^{m(s)}$, where $t = f(s)$.

The following proposition determines when S is an algebraic torus.

Proposition 4.4.3 *For* $e \in E$ *let* $m(e)$ *be the greatest common divisor of the numbers* $m(c)$ *for all* $c \in C$ *such that* $f(c) = e$. *Then* $\hat{S}$ *is torsion free if and only if the positive integers* $m(e)$ *for all* $e \in E$ *are pairwise coprime (if* E *has just one element,* $\hat{S}$ *is torsion free). This is the case if and only if* S *is an algebraic* k-*torus.*

Proof. Let $\gamma_e = m(e)^{-1} \sum_{f(c)=e} m(c)\gamma_c$. It is clear that the torsion subgroup of $\hat{S}$ is isomorphic to the torsion subgroup of $\bigoplus \mathbf{Z}\gamma_e / \langle m(e)\gamma_e - m(e')\gamma_{e'} \rangle$. This finite subgroup is non-trivial if and only if there exist integers n_e, $e \in E$, such that $\sum_{e \in E}(n_e/m(e)) = 0$ and $n_e/m(e) \notin \mathbf{Z}$ for some $e \in E$. The fact that this is possible if and only if $(m(e), m(e')) \neq 1$ for some $e \neq e'$ is almost immediate. QED

Corollary 4.4.4 *Suppose that every closed fibre of* f *(over points of* T*) contains a component of multiplicity 1. Then* S *is an algebraic* k-*torus.*

Let $\mathcal{T}$ be an X-torsor under S of type λ. By Theorem 4.3.1 the restriction of $\mathcal{T}$ to U_0, denoted by $\mathcal{T}_{U_0}$, is the pull-back $\phi^*(\mathcal{M}_R)$, where $\phi : U_0 \to R = \prod_{t \in T} R_{k_t/k}(\mathbf{G}_{m,k_t})$ is a map such that $\hat{\phi} : \hat{R} = \overline{k}[U_0']^* / \overline{k}^* \to \overline{k}[U_0]^*$ is a lifting of the natural injection $\overline{k}[U_0']^* \to \overline{k}[U_0]^*$.

Let $\alpha = \{\alpha_t\}$, where $\alpha_t \in k_t^*$, $t \in T$. Consider the map

$$\mathbf{A}^1_k \to \prod_{t \in T} R_{k_t/k}(\mathbf{A}_{m,k_t})$$

defined on $\overline{k}$-points by $x \mapsto \{\psi_{t,i}(\alpha_t)(x - e_{t,i}), t \in T, i = 1, \ldots, d_t\}$. Its restriction to U_0' maps this open set to $R = \prod_{t \in T} R_{k_t/k}(\mathbf{G}_{m,k_t})$. Let $\phi_\alpha : U_0 \to R$ be the composition of this map with f. Then ϕ_α for various α give all possible maps such that $\hat{\phi}_\alpha$ is a lifting of the natural injection $\rho = f^* : \overline{k}[U_0']^* / \overline{k}^* \to \overline{k}[U_0]^* / \overline{k}^*$. (We forget about the minus sign here.)

Let $V_\alpha = \phi_\alpha^*(\mathcal{M}_R)$. Then there exists $\alpha = \{\alpha_t\}$, $\alpha_t \in k_t^*$, $t \in T$, such that we have $\mathcal{T}_{U_0} = U_0 \times_{U_0'} V_\alpha$. Now we describe V_α by explicit equations and define its partial compactification. This geometric description will be used later in arithmetical applications.

Let z_s be a k_s-variable, for $s \in S$, and x and v the 'usual" k-variables. Then V_α can be given in

$$U_0' \times \mathbf{G}_{m,k} \times \prod_{s \in S} R_{k_s/k}(\mathbf{G}_{m,k_s}) \subset \mathbf{A}^1_k \times \mathbf{A}^1_k \times \prod_{s \in S} R_{k_s/k}(\mathbf{A}^1_{k_s})$$

by the following equations:

$$\alpha_t(x - a_t) = v^{-1} \prod_{t=f(s)} N_{k_s/k_t}(z_s)^{m(s)} \neq 0, \; z \neq 0, \; t \in T. \qquad (4.33)$$

Rewriting these as equations over k gives r equations in $n + 2$ variables, where $r = |E|$ and $n = |C|$. Consider the following partial compactification of V_α, the variety $W_\alpha \subset \mathbf{A}_k^1 \times \mathbf{A}_k^1 \times \prod_{s \in S} R_{k_s/k}(\mathbf{A}_{k_s}^1)$ given by

$$\alpha_t(u - a_t v) = \prod_{t=f(s)} N_{k_s/k_t}(z_s)^{m(s)}, \; (u, v) \neq (0, 0), \; t \in T, \qquad (4.34)$$

where $u = xz$. This variety is fibred over $\mathbf{P}_k^1$: we denote by ξ the morphism which sends a point of W_α to $(u : v) \in \mathbf{P}_k^1$. One sees that the fibres of ξ over the points of $U_0' = \mathbf{A}_k^1 \backslash T$ are torsors under S. We have $V_\alpha = \xi^{-1}(U_0')$. The complement $W_\alpha \backslash V_\alpha$ is a disjoint union of the fibres of ξ over the points $t \in T$ and the fibre at the point ∞ given by $(u : v) = (1 : 0)$. The fibre at ∞ is again a k-torsor under S (hence of dimension $n + 1 - r$). The fibre at a point t of T is the product of the scheme given by

$$\prod_{t=f(s)} N_{k_s/k_t}(z_s)^{m(s)} = 0,$$

and of the scheme

$$Y_{\alpha,t} \subset \mathbf{A}_k^1 \times \prod_{s \in S, f(s) \neq t} R_{k_s/k}(\mathbf{A}_{k_s}^1),$$

given by

$$\alpha_{t'}(a_t - a_{t'})v = \prod_{t'=f(s)} N_{k_s/k_{t'}}(z_s)^{m(s)}, \; v \neq 0, \; t' \in T \backslash \{t\}.$$

Let us observe that $Y_{\alpha,t}$ is a torsor under a k-group of multiplicative type S_t which is defined similarly to S by replacing T by $T \backslash \{t\}$. Then S_t is a quotient of S, in particular, if S is a torus, then the same is true for S_t. An easy computation shows that $W_\alpha \backslash V_\alpha$ is of equal dimension $n + 1 - r$.

Over $\overline{k}$ the equations (4.34) can be written as

$$\psi_{t,i}(\alpha_t)(u - e_{t,i}v) = \prod x_c^{m(c)}, \; (u, v) \neq (0, 0), \; t \in T, \; i = 1, \ldots, d_t, \qquad (4.35)$$

where the product is taken over $c \in C$ such that $f(c)$ is the point in E with coordinate $u/v = e_{t,i}$.

Lemma 4.4.5 *Assume that the characteristic of k is coprime with $m(e)$, for all $e \in E$ (these integers are defined in Proposition 4.4.3). Then the quasi-affine scheme W_α is smooth, geometrically integral, and contains V_α as a dense open set. The morphism $\xi : W_\alpha \to \mathbf{P}_k^1$ is faithfully flat.*

Proof. Without loss of generality we can assume that T is not empty. (Otherwise $W_\alpha = \mathbf{A}_k^2 \setminus (0,0)$, and there is nothing to prove.)

Let us show that the rank of the Jacobian matrix J of the equations (4.35) is maximal at any point of $\overline{W}_\alpha$. Suppose that a non-trivial linear combination of rows of J is 0. The rows of J consist of partial derivatives of the equations corresponding to the $e_{t,i}$'s. Since all the $e_{t,i}$'s are different, there must be at least three rows entering our linear combination with a non-zero coefficient. Now remark that each variable x_c enters just one equation of (4.35), namely the one corresponding to $e = f(c)$. Thus for each of these three e's we have an alternative: either the characteristic of k divides $m(c)$ for all c such that $e = f(c)$, or $x_c = 0$ for at least one c such that $e = f(c)$. In the first case the characteristic divides $m(e)$, thereby contradicting our assumption, hence this case does not occur. Therefore we have $x_c = x_{c'} = 0$ for some c and c' such that $f(c) \neq f(c')$. However, this implies that $(u, v) = (0, 0)$, which contradicts (4.35). Therefore the rank of the Jacobian matrix J is maximal at every point of $\overline{W}_\alpha$. In particular, W_α is smooth of equal dimension $n + 2 - r$.

We have seen that $W_\alpha \setminus V_\alpha$ is of equal dimension $n + 1 - r$, therefore V_α is dense in W_α.

Let us prove that $\overline{W}_\alpha$ is irreducible.

We start with the observation that the k-subgroup of $\mathbf{G}_{m,k}^n$ given by $x_1^{m_1} \ldots x_s^{m_s} = 1$, $m_1, \ldots, m_s \in \mathbf{Z}$, is an extension of μ_m by a torus, where m is the greatest common divisor of $m_1, \ldots, m_s$. (This is equivalent to the fact that $(\mathbf{Z}^s / \langle (m_1, \ldots, m_s) \rangle)_{tors} \simeq \mathbf{Z}/m$.) Suppose that the characteristic of k does not divide m, and that the m-th roots of 1 are in k. Then it follows that if $a \in k^*$ is not an m-th power in k, then the k-variety $x_1^{m_1} \ldots x_s^{m_s} = a$ is integral. The integral closure of k in the field of functions of this variety, the field $k(a^{1/m})$, is a cyclic Galois extension of k.

Let us apply this to the study of $\overline{W}_\alpha$. Looking at the equations (4.35) one observes that the morphism $\overline{W}_\alpha \to \mathbf{A}_{\overline{k}}^2 \setminus \{(0,0)\}$ given by (u, v) represents $\overline{W}_\alpha$ as the fibre product of the irreducible (rational) $\overline{k}$-varieties X_e given by $u - x(e)v = \prod_{f(c)=e} x_c^{m(c)}$, $(u, v) \neq (0, 0)$, over $\mathbf{A}_{\overline{k}}^2 \setminus \{(0,0)\}$. Let $K = \overline{k}(u, v)$, and let $X_{e,K}$ be the generic fibre of $X_e \to \mathbf{A}_{\overline{k}}^2 \setminus \{(0,0)\}$. This is the variety of the type considered above: $X_{e,K}$ is integral, and the integral closure of K in the field of functions $\overline{k}(X_e)$ is $K_e = \overline{k}(u,v)((u - x(e)v)^{1/m(e)})$.

Note that the extensions K_e/K are linearly disjoint, and their compositum L/K is Galois with the Galois group $\prod_{e \in E}(\mathbf{Z}/m(e))$ (indeed, the functions $u - x(e)v$ are integrally linearly independent in $\overline{k}(u,v)^*$). Therefore the fibre of $\overline{W_\alpha}$ over $Spec(K)$, which is $\prod_{e \in E} X_{e,K}$, is integral.

This proves that only one irreducible component of $\overline{W_\alpha}$ is dominant over $\mathbf{A}_{\overline{k}}^2 \setminus \{(0,0)\}$. Since $\overline{W_\alpha}$ is of equal dimension $n + 2 - r$, and the fibres of $\overline{W_\alpha} \to \mathbf{A}_{\overline{k}}^2 \setminus \{(0,0)\}$ have equal dimension $n - r$, every irreducible component must be dominant over $\mathbf{A}_{\overline{k}}^2 \setminus \{(0,0)\}$. Hence $\overline{W_\alpha}$ is irreducible.

Finally, we note that since W_α dominates $\mathbf{P}_k^1$, the morphism ξ is flat ([Hartshorne], III.9.7). In fact, ξ is surjective and thus is faithfully flat. QED

Corollary 4.4.6 *A dense open subset of an X-torsor associated to the morphism $f : X \to \mathbf{P}_k^1$ (in the notation of this section, of any X-torsor of type λ) is isomorphic to a dense open subset of the fibre product $X \times_{\mathbf{P}_k^1} W_\alpha$, taken with respect to $f : X \to \mathbf{P}_k^1$ and $\xi : W_\alpha \to \mathbf{P}_k^1$, for some α. The converse is also true: for any $\alpha = \{\alpha_t\}$, $\alpha_t \in k_t^*$, a dense open subset of $X \times_{\mathbf{P}_k^1} W_\alpha$ is isomorphic to a dense open subset of an X-torsor of type λ.*

This follows from Lemma 4.4.5 which asserts that W_α contains V_α as a dense open set, and from Theorem 4.3.1.

Example 1.

Suppose that $|E| = 2$. Then the morphism which forgets u and v, maps W_α isomorphically onto an open subset of $\mathbf{A}_k^n$ whose complement has codimension 2. (Indeed, we can verify the statement over $\overline{k}$, where it immediately follows from (4.35).) If $|E| = 3$, the fibres X_t are integral, and $[k_s : k_t] = 2$ for all t, then W_α is a dense open subset of a quadric.

Example 2: explicit 2-descent on hyperelliptic curves.

Suppose that X_a is a smooth proper curve of genus g represented as a double covering $f : X_a \to \mathbf{P}_k^1$ given by

$$y^2 = aP(x),$$

where $a \in k^*$, and $P(x)$ is a separable monic polynomial of even degree $n = 2 + 2g$ (we assume that the characteristic of k is different from 2).

We choose T as the ramification locus of f (given in $\mathbf{A}_k^1$ by $P(x) = 0$). It follows from the definition of S that

$$\hat{S} = (\bigoplus_{e \in E} \mathbf{Z}\gamma_e)/\langle 2\gamma_e - 2\gamma_{e'} \rangle.$$

Thus $\hat{S}$ is an extension of the trivial Γ_k-module $\mathbf{Z}$ by $\hat{S}_{tors}$. Next, $\hat{S}_{tors}$ is the submodule of the permutation module $\bigoplus_{e \in E} \mathbf{Z}/2$ given by the elements with zero sum. The submodule $Ker(\lambda) \simeq \mathbf{Z}/2 \subset \hat{S}$ is generated by $(1, 1, \ldots , 1) - (n, 0, \ldots , 0)$. This follows from the well known fact that the 2-torsion in the Jacobian of a hyperelliptic curve is generated by the differences of its Weierstrass points, with relations given by the divisors of the functions $(x - x(e))(x - x(e'))^{-1}$, $e, e' \in E$, $e \neq e'$, and the divisor of the function $y(x - x(e))^{-n/2}$, for some $e \in E$ (see, e.g., [Mumford, T], p. 3.30–3.32). This implies that $\overline{k}[U]^*$ is generated by $\overline{k}[U']^*$ and the function $y(x - x(e))^{-n/2}$. The divisor of the last function is precisely the above generator of $Ker(\lambda)$.

Consider the k-group of multiplicative type S' such that $\hat{S}' = \hat{S}/Ker(\lambda)$. In other terms

$$\hat{S}' = Div_{\overline{X_a} \setminus \overline{U}}(\overline{X_a})/(\overline{k}[U]^*/\overline{k}^*).$$

In a similar way, we define a finite k-group S'' so that $\hat{S}'' = \hat{S}_{tors}/Ker(\lambda)$. The sets of $\overline{k}$-points of these k-groups are

$$
\begin{aligned}
S(\overline{k}) &= \{(\epsilon_1 s, \ldots , \epsilon_n s), \ s \in \overline{k}^*, \ \epsilon_i = \pm 1\}, \\
S'(\overline{k}) &= \{(\epsilon_1 s, \ldots , \epsilon_n s), \ s \in \overline{k}^*, \ \epsilon_i = \pm 1, \ \textstyle\prod \epsilon_i = 1\}, \\
S''(\overline{k}) &= \{(\epsilon_1, \ldots , \epsilon_n), \ \epsilon_i = \pm 1, \ \textstyle\prod \epsilon_i = 1\}/\langle(-1, \ldots , -1)\rangle.
\end{aligned}
$$

Let $\lambda' : \hat{S}' \to Pic(\overline{X_a})$ be the homomorphism induced by λ. Then $\lambda'(\hat{S}')$ is the group of divisors generated by the Weierstrass points. Let $\lambda'' : \hat{S}'' \to Pic(\overline{X_a})$ be the restriction of λ'. The image $\lambda''(\hat{S}'')$ is generated by differences of Weierstrass points. As a finite group $\hat{S}'' = (\mathbf{Z}/2)^{n-2} = (\mathbf{Z}/2)^{2g}$, and λ'' is a natural isomorphism of the Γ_k-module $\hat{S}''$ with $Jac(\overline{X_a})[2]$; $\lambda''(\gamma_e - \gamma_{e'})$ is the divisor $(x(e), 0) - (x(e'), 0)$. In the notation of Section 3.3 we have $\lambda'' = \lambda_2$. Recall that an $\overline{X_a}$-torsor of this type is obtained by restricting to $\overline{X_a} \subset Jac(\overline{X_a})$ the torsor over $Jac(\overline{X_a})$ given by the isogeny $x \mapsto 2x$.

Let K be a k-algebra $K = k[x]/(P(x)) = \prod_{t \in T} k_t$, $\alpha = \{\alpha_t\} \in K^*$, and $N(\alpha) = \prod_{t \in T} N_{k_t/k}(\alpha_t)$. Equations (4.34) can now be written as

$$\alpha_t(u - a_t v) = z_t^2, \ (u, v) \neq (0, 0), \ t \in T. \tag{4.36}$$

Over $\overline{k}$ this can be written as

$$\alpha_i(u - e_i v) = z_i^2, \ (u, v) \neq (0, 0), \ i = 1, \ldots , n, \tag{4.37}$$

where $\alpha_i \in \overline{k}^*$, and $e_i \in \overline{k}$ are such that $e_i \neq e_j$ if $i \neq j$. Eliminating u and v we see that W_α is the punctured affine cone over a smooth projective

curve Y_α which is a complete intersection of $n-2$ quadrics in $\mathbf{P}_k^{n-1}$. By
Corollary 4.4.6 any X_a-torsor $\mathcal{T}$ of type λ contains a dense open subset
isomorphic to a dense open subset of $X_a \times_{\mathbf{P}_k^1} W_\alpha$. On taking norms and
multiplying, we see from equations (4.36) that $P(u/v) = N(\alpha)^{-1} z^2$, where
$z = v^{-n/2} \prod_t N_{k_t/k}(z_t) \in k(W_\alpha)^*$. Thus a dense open subset of $\mathcal{T}$ is
isomorphic to a dense open subset of

$$W_\alpha \times_k Spec(k(\sqrt{a^{-1} N(\alpha)})).$$

The fact that this is not geometrically irreducible agrees with the fact that
$Ker(\lambda) = \mathbf{Z}/2$ (cf. Exercise 2 of Chapter 2).

Proposition 4.4.7 *The map sending* $(u, v; z_t)$ *to*

$$x = v^{-1} u, \quad y = v^{-n/2} \prod_{t \in T} N_{k_t/k}(z_t)$$

defines a morphism

$$\tau : Y_\alpha \to X_{N(\alpha)}.$$

This morphism makes Y_α *an* $X_{N(\alpha)}$*-torsor of type* $\lambda'' = \lambda_2$.

Note that the morphism $\xi : W_\alpha \to \mathbf{P}_k^1$ factors as

$$\xi : W_\alpha \to Y_\alpha \xrightarrow{\tau} X_{N(\alpha)} \xrightarrow{f} \mathbf{P}_k^1,$$

where the first map is an obvious torsor under $\mathbf{G}_m$.

Proof of proposition. To see that $\tau : Y_\alpha \to X_{N(\alpha)}$ is a torsor is easy: it is
enough to check this over $\overline{k}$, and for this it is enough to verify that the finite
group $S''(\overline{k})$ acts freely on $\overline{Y_\alpha}$. This easily follows from the presentation of
$S''(\overline{k})$ given above and equations (4.37). To prove that this is a torsor of
type λ'' one can proceed as follows.

By the last statement of Corollary 4.4.6 to α there corresponds an $X_{N(\alpha)}$-
torsor $\mathcal{T}$ under S of type λ such that a dense open subset of $\mathcal{T}$ is a dense
open subset of $X_{N(\alpha)} \times_{\mathbf{P}_k^1} W_\alpha$. The exact sequence

$$1 \to S' \to S \to \mathbf{Z}/2 \to 0$$

gives rise to

$$H^1(X, S') \to H^1(X, S) \to H^1(X, \mathbf{Z}/2).$$

The right arrow sends the class of $\mathcal{T}$ to the class of the quotient (the push-
forward) $\mathcal{T}/S'$. Its dense open subset is isomorphic to a dense open subset
of

$$X_{N(\alpha)} \times_{\mathbf{P}_k^1} W_\alpha/S' = X_{N(\alpha)} \times_{\mathbf{P}_k^1} X_{N(\alpha)}.$$

The normalization of this curve is the union of two disjoint copies of $X_{N(\alpha)}$. This implies that the $X_{N(\alpha)}$-torsor $\mathcal{T}/S'$ under $\mathbf{Z}/2$ is trivial. Therefore $\mathcal{T}$ is a push-forward of an $X_{N(\alpha)}$-torsor under S' of type λ', say $\mathcal{T}'$. Then, according to the discussion before the proposition, a dense open subset of $\mathcal{T}$ is a dense open subset of the union of two disjoint copies of W_α, and on the other hand, it is isomorphic to a dense open subset of the union of two disjoint copies of $\mathcal{T}'$. Since W_α is geometrically integral by the previous lemma, $\mathcal{T}'$ is birationally equivalent to W_α. Passing to quotients by $\mathbf{G}_m$ we obtain that $\mathcal{T}'/\mathbf{G}_m$, which is an $X_{N(\alpha)}$-torsor under S'' of type λ'', is birationally equivalent to Y_α, hence it is isomorphic to Y_α (since birationally equivalent smooth projective curves are isomorphic). Thus Y_α is indeed an $X_{N(\alpha)}$-torsor of type λ''. QED

This generalizes Proposition 3.3.6 *(b)* to arbitrary hyperelliptic curves, and provides a sufficient condition for the existence of torsors of type λ_2 on them.

Actually, one does not need the difficult Theorem 4.3.1 to prove Proposition 4.4.7. (Although that general result is indispensable in many other applications.) For example, to prove directly that $Im(\mathtt{type}(\tau)) = Jac(\overline{X_a})[2]$ one can argue as follows. By (2.5) the image of $\mathtt{type}(\tau)$ is the kernel of $\tau^* : Pic(\overline{X_a}) \to Pic(\overline{Y_\alpha})$ (cf. the discussion at the beginning of Section 2.3). But with the notation of equations (4.37) it is easy to see that the inverse image in Y_α of the Weierstrass point of X_a with coordinates $(e_i, 0)$ is the hyperplane section $z_i = 0$. Hence $\tau^{-1}((e_i, 0) - (e_j, 0))$ is the divisor of the function z_i/z_j. This means that differences of Weierstrass points become principal divisors on Y_α, therefore $Jac(\overline{X_a})[2] \subset Ker(\tau^*)$. Comparing the orders of these groups we conclude that this must be an equality.

Example 3: explicit descent on conic bundle surfaces.

If S is an algebraic k-torus, then W_α is $\overline{k}$-rational. Indeed, in this case $\overline{V_\alpha}$ is a torsor over $\overline{U'_0} \subset \mathbf{A}^1_{\overline{k}}$ under $\mathbf{G}_m^{n+1-r}$. We have $H^1(\overline{U'_0}, \mathbf{G}_m) = Pic(\overline{U'_0}) = 0$, thus $\overline{V_\alpha} \cong \overline{U'_0} \times \mathbf{G}_m^{n+1-r}$, and so $\overline{V_\alpha}$ is rational. Hence W_α, which by Lemma 4.4.5 contains V_α as a dense open set, is $\overline{k}$-rational.

Such is the case when the generic fibre of f is geometrically integral, and f everywhere has a local section for the étale topology. For example, $f : X \to \mathbf{P}^1_k$ is a conic bundle surface or a pencil of curves of genus 1 without multiple fibres. In this situation, one can take T to consist of the closed points of $\mathbf{P}^1_k$ corresponding to the fibres which are not geometrically integral. For arithmetical purposes a more economical choice of T

suffices: T should consist of the closed points whose fibres do not contain an irreducible component s of multiplicity 1 such that $k_s = k_t$.

As a more detailed example we consider a smooth and proper conic bundle surface X_a given by the equation (see Exercise 3 below for a detailed construction of X_a)

$$y^2 - bz^2 = aP(x), \qquad (4.38)$$

where $a, b \in k^*$, and $P(x) \in k[x]$ is separable, monic, and of even degree n. (We continue to assume that the characteristic of k is different from 2.) As in Example 2 the morphism $f : X_a \to \mathbf{P}_k^1$ is given by the projection to the x-axis. We keep the notation of Example 2. The difference from Example 2 is that now S is a torus, and λ is injective. The image $\lambda(\hat{S})$ is generated by the classes of the lines $y = \pm\sqrt{b}z$, $x = x(e)$, $e \in E$.

Proposition 4.4.8 *The X_a-torsor of type λ under S defined by $\alpha_t \in K^*$ is birationally equivalent to the product of $\mathbf{A}_k^1$, the conic $y^2 - bz^2 = aN_{K/k}(\alpha)$, and the geometrically integral complete intersection Y_α of $n - 2$ quadrics in $\mathbf{P}_k^{2n-1}$ obtained by eliminating u and v from the equations*

$$\alpha_t(u - a_t v) = x_t^2 - by_t^2, \ (u,v) \neq (0,0), \ t \in T, \qquad (4.39)$$

where x_t and y_t are variables with values in k_t.

Proof. It is clear that W_α is birationally equivalent to $Y_\alpha \times_k \mathbf{A}_k^1$ (since W_α is the punctured affine cone over a dense open subset of Y_α). By virtue of Corollary 4.4.6 it will be enough to show that over the field of functions $k(Y_\alpha)$ the conic $y^2 - bz^2 = aP(u/v)$ is isomorphic to the conic $y^2 - bz^2 = aN_{K/k}(\alpha)$. We have

$$N_{k_t/k}(x_t^2 - by_t^2) = N_{k_t/k}N_{k_t(\sqrt{b})/k_t}(x_t + \sqrt{b}y_t)$$
$$= N_{k(\sqrt{b})/k}N_{k_t(\sqrt{b})/k(\sqrt{b})}(x_t + \sqrt{b}y_t).$$

Taking the product for all $t \in T$ we obtain from (4.39) that

$$N(\alpha)P(u/v) = v^{-n} \prod_{t \in T} N_{k_t/k}(x_t^2 - by_t^2)$$

is a norm of $v^{-n/2}\prod_{t \in T} N_{k_t(\sqrt{b})/k(\sqrt{b})}(x_t + \sqrt{b}y_t)$ for the quadratic extension $k(Y_\alpha)(\sqrt{b})/k(Y_\alpha)$. This proves what we want. QED

For future arithmetic applications we observe that Y_α contains a pair of skew $\mathbf{P}^{n-1}$'s which are conjugate over $k(\sqrt{b})$. For example, take $x_t = \pm\sqrt{b}y_t$ for all $t \in T$.

See Exercise 4 below for a descent on a non-proper conic bundle surface

Exercises

1. Suppose that the characteristic of the field K is not 2. For $\alpha, \beta \in K^*$ we denote by (α, β) the quaternion algebra $K + Ki + Kj + Kij$ subject to the standard relations $i^2 = \alpha$, $j^2 = \beta$, $ij = -ji$. We denote by $\langle \alpha, \beta \rangle$ its class in $\mathrm{Br}(K)$. Let an elliptic curve E be given by the equation $y^2 = (x - c_1)(x - c_2)(x - c_3)$, where $c_i \in K$, and $c_i \neq c_j$ for $i \neq j$. Show that for $a \in K^*$ the element $\langle a, x - c_i \rangle \in \mathrm{Br}(K(E))$ is *unramified*, that is, belongs to the image of the (injective) map $\mathrm{Br}(E) \to \mathrm{Br}(K(E))$. (For this it is enough to show that E can be covered by open sets U_i such that for each i our element can be represented by the quaternion algebra (α_i, β_i) where $\alpha_i, \beta_i \in K[U_i]^*$.)

2. We keep the notation of Exercise 1. Using the Example in Section 4.1 show that every element of $\mathrm{Br}(E)$ of order 2 which vanishes at ∞ (the neutral element of the group law on E) is of the form $\langle a, x - c_1 \rangle + \langle b, x - c_2 \rangle$ for some $a, b \in K^*$. Show that this element is 0 if and only if the class of (a, b) in $(K^*/K^{*2})^2$ is the image of a K-point of E under the Kummer map $E(K)/2 \to H^1(k, E[2])$. If we identify $E[2]$ with $(\mathbf{Z}/2)^2$ with the basis $(c_1, 0)$, $(c_2, 0)$, then the Kummer map sends (X, Y) to $(X - c_1, X - c_2)$, when $X \neq c_1$, $X \neq c_2$, and sends $(c_1, 0)$ (resp. $(c_2, 0)$, resp. ∞) to $((c_1 - c_2)(c_1 - c_3), c_1 - c_2)$ (resp. to $(c_2 - c_1, (c_2 - c_1)(c_2 - c_3))$, resp. to $(1, 1))$.

3. Build a natural smooth proper model of the affine surface in $\mathbf{A}^3_k$ given by (4.38) as follows. Let $y^2 - bz^2 = aP(x)t^2$ be a surface in $\mathbf{P}^2_k \times_k \mathbf{A}^1_k$, and let $Y^2 - bZ^2 = au^n P(u^{-1})T^2$ be another such surface. Then X_a is obtained by gluing together these two surfaces over $\mathbf{P}^1_k \setminus \{0, \infty\}$ by $x = u^{-1}$, $y = Y$, $z = Z$, $t = u^{n/2}T$. Observe that the lines l, l' given by $y = \pm\sqrt{b}z$, $t = 0$, are disjoint conjugate sections of $X_a \to \mathbf{P}^1_k$.

4. In the notation of Example 3 above let $\alpha \in K^*$, $a = N_{K/k}(\alpha)$. Show that if x, y, z are defined by

$$x = v^{-1}u, \quad y \pm \sqrt{b}x = v^{-n/2} \prod_{t \in T} N_{k_t(\sqrt{b})/k(\sqrt{b})}(x_t \pm \sqrt{b}y_t),$$

then $y^2 - bz^2 = N_{K/k}(\alpha)P(x)$. Check that this defines a morphism $W_\alpha \to X_a \setminus (l \cup l')$. Prove that this is a universal torsor.

Comments

The material of the first section is taken from [S99]. This is an essential albeit technical improvement of the methods of [CS87a] which are applicable only to torsors under tori. This is a key technical point which connects the descent obstruction related to torsors under groups of multiplicative type with the Manin obstruction provided by the Brauer group.

The main result of the second section is the commutative diagram (4.15) which has surprisingly many applications. (The one not discussed in this book is the computation of the Brauer group of elliptic surfaces in the final section of [CSS98b].) Armed with these technicalities we give a complete (and partly new) proof of the local description of torsors ([CS87a], 2.3). (The use of derived categories allows us to avoid referring to the rather involved homological algebra of [CS87a], Sect. 1.)

The usefulness of this description is demonstrated in the examples in the last section. It should be said here that it was Per Salberger who first wrote the equations like (4.34) in the context of his approach to conic bundles via K-theory of orders (*Sur l'arithmétique de certaines surfaces de del Pezzo.* C. R. Acad. Sci. Paris **303** (1986) 273–276). It was quickly realized by Colliot-Thélène and Sansuc, who were also interested in conic bundles, that these equations describe torsors of the type considered above, which for conic bundles are particularly close to universal torsors (*La descente sur les surfaces rationnelles fibrées en coniques.* C. R. Acad. Sci. Paris **303** (1986) 303–306). These ideas were applied to the arithmetic of various fibrations over $\mathbf{P}^1$: fibrations into quadrics of relative dimension 2 ([S90a], recently improved in [CS00]; these works build on the observations of Example 1 above), to conic bundles with four [C90] and five degenerate fibres [SS].

To those working with descent it has been clear that these equations resemble strikingly the equations for 2- or 4-descent of elliptic curves, the only difference being that squares are replaced by norms. Explicit descent on curves, in particular, 2-descent on hyperelliptic curves, was studied by Cassels and other mathematicians in numerous papers, mainly aimed at developing effective methods for computing the rational points on such curves and the rank of their Jacobians (see [CFl] and references therein). Our exposition, which subsumes the geometry of these various descents, is a somewhat generalized version of [S96].

It was hoped that the introduction of universal torsors could lead to the understanding of the Hasse principle and other arithmetic properties of rational varieties, say cubic surfaces. The main obstacle, however, is that the universal torsors (or torsors closely related to them) are usually given by complicated equations. For example, in the case of diagonal cubic surfaces one is led to specific singular intersections of two cubics in $\mathbf{P}^9$ (see [CS87a], 2.5, and [CKS], Prop. 11). It seems to us that a conceptual approach to these equations is not yet known.

The quaternion algebras considered in Exercises 1 and 2 were used in [YM], [GMY] to construct explicit generators of the 2-torsion subgroup of the Brauer group of an elliptic curve defined over a local field.

Part two

DESCENT AND MANIN OBSTRUCTION

Notation. We keep the notation of Part one.

In this part, unless otherwise stated, k is a number field; Ω is the set of places of k; Ω_∞ is the set of archimedean places; $\mathcal{O}_k$ is the ring of integers of k; when $S \subset \Omega$ we denote by $\mathcal{O}_{k,S}$ the ring of S-integers of k (integers outside S); k_v is the completion of k with respect to the place $v \in \Omega$; $\mathcal{O}_v$ is the ring of integers of k_v; $\mathbb{A}_k$ is the adèle ring of k. If Σ is a subset of Ω, then $\mathbb{A}_k^\Sigma$ stands for the adèles without v-components for $v \in \Sigma$, $k_\Sigma = \prod_{v \in \Sigma} k_v$, $k^\Sigma = \prod_{v \in \Omega \setminus \Sigma} k_v$. If M is a discrete Γ_k-module, we write

$$\text{III}^i(k, M) := Ker\big[H^i(k, M) \longrightarrow \prod_{v \in \Omega} H^i(k_v, M)\big].$$

If A is an abelian variety over k, then $\text{III}(A)$ is its *Tate–Shafarevich group*, $\text{III}(A) = \text{III}^1(k, A(\overline{k}))$.

If X is a variety over a number field k, we write $X_v = X \times_k k_v$.

Brauer groups. If X is a scheme over the base Y, then for any scheme Y'/Y the pairing

$$Br(X) \times X(Y') \to Br(Y'), \quad (A, s) \longmapsto A(s)$$

is obtained by specializing elements of $Br(X)$: if we are given a Y-morphism $s : Y' \to X$, then $A(s)$ is the pull-back $s^*(A) \in Br(Y')$. If X is a smooth geometrically integral variety over a field k of characteristic 0, then

$$Br_{nr}(X) = Br_{nr}(k(X)/k) \subset Br(X)$$

denotes the *unramified Brauer group* of the field $k(X)$ over k. It can be identified with the Brauer group of any smooth proper k-variety birationally equivalent to X (see [Grothendieck 68]).

The *local invariant* of the Brauer group of a local field k_v is a homomorphism $\text{inv}_v : Br(k_v) \hookrightarrow \mathbf{Q}/\mathbf{Z}$. It is an isomorphism for finite v. If v is a real place, inv_v identifies $Br(k_v)$ with $\frac{1}{2}\mathbf{Z}/\mathbf{Z}$, and when v is complex, $Br(k_v)$ is 0.

For a subset Σ of Ω we denote by $Br^\Sigma(X)$ the subgroup of $Br_1(X)$ consisting of elements A whose image in $Br_1(X_v)$ comes from $Br(k_v)$ for all $v \in \Sigma$. We have $Br(X)^\emptyset = Br_1(X)$, and we set

$$\text{Б}(X) = Br^\Omega(X).$$

We also define $Br_{nr}^\Sigma(X/k) = Br^\Sigma(X) \cap Br_{nr}(X)$.

Obstructions over number fields

In this chapter we discuss the Hasse principle and various approximation properties for varieties defined over number fields, as well as known obstructions to them.

The short survey of known results on the Hasse principle and weak approximation in the first section is just an introduction to the main object of interest here: obstructions and their interrelations. Fortunately, this subject is covered in a few excellent survey articles [Sansuc 82], [MT], [Sansuc 87], [C87], [C92], [SD96], [C98]. The Manin obstruction to the Hasse principle, and its various Ramifications, are defined in the second section.

In the last setion we discuss the obstructions to the Hasse principle and weak approximation obtained via descent with torsors under (possibly, noncommutative) algebraic groups. Examples (Chapter 8) show that these obstructions can be finer than the Manin obstruction. The classical theory of descent on elliptic curves and their principal homogeneous spaces worked with torsors under finite abelian (algebraic) groups. This is covered by the Manin obstruction. To go beyond it one applies the same idea in the nonabelian setting.

5.1 The Hasse principle, weak and strong approximation

A class of geometrically integral varieties over a number field k satisfies *the Hasse principle* if for every variety in this class the condition $X(k_v) \neq \emptyset$ for all places v of k implies $X(k) \neq \emptyset$. Usually we shall speak of the Hasse principle for smooth varieties. If the Hasse principle holds for the smooth loci of varieties in a certain class, we shall say that this class satisfies the *smooth Hasse principle*.

For proper varieties we have $X(k_\Omega) = X(\mathbb{A}_k)$. Indeed, if X is projective, then given a solution of a system of homogeneous equations we can get rid of coefficients' denominators. For a non-archimedean field k_v this means that the set $X(k_v)$ is the same as the set of solutions with coefficients in the ring of integers $\mathcal{O}_v$. If X is proper, then, more technically, we can find a proper scheme $\mathcal{X} \to Spec(\mathcal{O}_{k,S})$, for some finite $S \subset \Omega$, with generic fibre X. Then for the primes $v \notin S$ we have $\mathcal{X}(\mathcal{O}_v) = X(k_v)$ by the valuative criterion of properness. This implies $X(k_\Omega) = X(\mathbb{A}_k)$.

Let us list without proof some of the classical and more recent results on the Hasse principle. Speaking of cubics and intersections of two quadrics we shall always assume that they are geometrically integral, non-conical (cannot be reduced to a lesser number of variables by a linear transformation), and of codimension 1 and 2, respectively.

Theorem 5.1.1 *The following classes of geometrically integral varieties over a number field k satisfy the Hasse principle:*

(a) smooth projective quadrics (Minkowski and Hasse);

(b) Severi–Brauer varieties, that is, $(\overline{k}/k)$-forms of $\mathbf{P}^n_k$ (Châtelet);

(c) smooth projective cubics in $\mathbf{P}^n_k$, $k = \mathbf{Q}$, $n \geq 9$ (Hooley);

(d) smooth proper models of the intersections of two quadrics in $\mathbf{P}^n_k$, $n \geq 8$ ([CSS87]);

(e) principal homogeneous spaces under simply connected, or adjoint, semisimple groups (Kneser, Harder and Chernousov).

This list is very far from being complete. Rather far-reaching results are known for cubic hypersurfaces, complete intersections of two quadrics, fibrations into quadrics, etc. The analysis may become simpler if such a variety contains a specific configuration of singular points (see [CSS87], [CS89], [S90a], [H95], [CLSS]). If one considers complete intersections of some fixed multidegree, then one can usually prove by analytic methods that, at least over $k = \mathbf{Q}$, the smooth Hasse principle holds provided the dimension is large compared with the dimension of the set of singular points, cf. [Birch]. We refer the reader to the survey papers [Sansuc 82], [MT], [C92], [SD96], [C98]. A complete proof of *(e)* can be found in [PR], VI.

Counter-examples to the Hasse principle will be discussed in subsequent chapters.

Sometimes, one can prove that the Hasse principle holds by showing first that the Manin obstruction to it is the only one, and then computing that $\mathrm{Br}(X)$ reduces to $\mathrm{Br}(k)$. We shall continue this discussion after defining the Manin obstruction in Section 5.2.

A geometrically integral, smooth variety X satisfies *weak approximation* if the diagonal image of $X(k)$ is dense in $X(k_\Omega)$ in the product topology. Equivalently, X satisfies weak approximation if $X(k)$ is dense in $\prod_{v \in S} X(k_v)$ for any finite set of places $S \subset \Omega$. If Σ is a subset of Ω, then X satisfies weak approximation *away from* Σ if $X(k)$ is dense in $X(k^\Sigma)$. It is important to stress that weak approximation so defined implies that X satisfies the Hasse principle.

Theorem 5.1.2 *The following geometrically integral smooth varieties over a number field k satisfy weak approximation:*

   (a) k-rational varieties;

   (b) simply connected, adjoint, or almost simple, semisimple k-groups;

   (c) smooth projective intersections of two quadrics with a k-point in $\mathbf{P}^n_k$, $n \geq 6$.

The proof of *(a)* is easily obtained by reduction to the case of the affine line (draw a k-rational curve close to a given collection of local points), where it is the usual weak approximation theorem (applicable to any finite system of independent metrics on a field). A proof of *(b)* can be found in ([PR], VII.7.3, Thm. 8, Prop. 11). Any connected algebraic group satisfies the weak approximation property outside a certain set of finite places (see [PR], VII.7.3, Thm. 7). The statement *(c)* is easy ([CSS87], 3.11). We could repeat here the remarks we made about the Hasse principle; the proof of weak approximation is sometimes obtained along the same lines as the proof of the Hasse principle.

For some classes of varieties an approximation property stronger than weak approximation is available. Let Σ be a subset of Ω. One says that *strong approximation* holds away from Σ if the diagonal image of $X(k)$ is dense in $X(\mathbb{A}_k^\Sigma)$. (Recall that $\mathbb{A}_k^\Sigma$ are the adèles with v-components removed for $v \in \Sigma$.) Note that for proper varieties there is no difference between strong and weak approximation, but this is far from being true in general. For instance, strong approximation does not hold for $X = \mathbf{A}_k^1$ and $\Sigma = \emptyset$, but holds for this variety whenever Σ is not empty (k being discrete in $\mathbb{A}_k$).

For a connected algebraic group G we have the following necessary and sufficient condition for strong approximation away from a *non-empty* finite set of places $\Sigma \subset \Omega$: the Levi subgroup (the reductive part) of G must be simply connected and semisimple and contain no k-simple component G' such that $G'(k_\Sigma)$ is compact (Kneser and Platonov; see [PR], VII.7.4, Thm. 12). Strong approximation never holds for a variety with a non-trivial fundamental group (Minchev).

In this book we concentrate our attention on the Hasse principle, and various approximation properties will play a somewhat auxiliary rôle.

5.2 The Manin obstruction

The idea of the Manin obstruction is based on the following fundamental fact from the global class field theory. The *Hasse reciprocity law* states that the following sequence of abelian groups is exact:

$$0 \to \mathrm{Br}(k) \to \sum_{v \in \Omega} \mathrm{Br}(k_v) \to \mathbf{Q}/\mathbf{Z} \to 0.$$

Here the second arrow is the natural diagonal map, and the third arrow is the sum of *local invariants* $\mathrm{inv}_v : \mathrm{Br}(k_v) \hookrightarrow \mathbf{Q}/\mathbf{Z}$.

Let X be a smooth and geometrically integral variety over a number field k. Following Manin we consider the pairing

$$\mathrm{Br}(X) \times X(\mathbb{A}_k) \to \mathbf{Q}/\mathbf{Z}, \quad (A, \{P_v\}) \longmapsto \sum_{v \in \Omega} \mathrm{inv}_v(A(P_v)). \qquad (5.1)$$

We shall call this *the adelic Brauer–Manin pairing*.

The sum in the formula is finite. Indeed, there exists a finite subset $\Sigma \subset \Omega$ such that X extends to a scheme $\mathcal{X}$ over $Spec(\mathcal{O}_{k,\Sigma})$, and A extends to an element of $\mathrm{Br}(\mathcal{X})$. We can assume that Σ is big enough so that the v-components of our adelic point are in $\mathcal{X}(\mathcal{O}_v)$ for $v \notin \Sigma$. A point $P_v \in \mathcal{X}(\mathcal{O}_v)$ can be interpreted as a local section of $\mathcal{X} \to Spec(\mathcal{O}_{k,\Sigma})$ at $Spec(\mathcal{O}_v)$. The specialization of A at $P_v \in \mathcal{X}(\mathcal{O}_v)$ is an element of $\mathrm{Br}(\mathcal{O}_v)$. Since $\mathrm{Br}(\mathcal{O}_v)$ is trivial (see [Milne, EC], IV.1) we obtain $\mathrm{inv}_v(A(P_v)) = 0$ for all $v \notin \Sigma$.

Another property of this pairing is that it is trivial on elements of $\mathrm{Br}_0(X)$ (by the Hasse reciprocity law), so that it can be considered as a pairing between $\mathrm{Br}(X)/\mathrm{Br}_0(X)$ and $X(\mathbb{A}_k)$.

There is another variant of this pairing:

$$\mathrm{Br}_{nr}(X) \times X(k_\Omega) \to \mathbf{Q}/\mathbf{Z}, \quad (A, \{P_v\}) \longmapsto \sum_{v \in \Omega} \mathrm{inv}_v(A(P_v)), \qquad (5.2)$$

which we shall call *the product Brauer–Manin pairing*. It coincides with the previous one for proper varieties X, because then $X(\mathbb{A}_k) = X(k_\Omega)$ and $\mathrm{Br}_{nr}(X) = \mathrm{Br}(X)$. Let us show that the sum in (5.2) is finite. Let X_c be a smooth proper variety containing X as a dense open subset (it exists by Hironaka's theorem). The elements of $\mathrm{Br}_{nr}(X)$ uniquely extend to elements of $\mathrm{Br}(X_c)$. On the other hand, $X(k_\Omega) \subset X_c(k_\Omega) = X_c(\mathbb{A}_k)$. Therefore the finiteness of the sum in (5.2) follows from the finiteness of the sum in (5.1).

Let us define $X(\mathbb{A}_k)^{\mathrm{Br}(X)}$ and $X(k_\Omega)^{\mathrm{Br}_{nr}(X)}$ as 'the right kernels' of these pairings, that is, the subsets of $X(\mathbb{A}_k)$ (resp. of $X(k_\Omega)$) orthogonal to all elements of $\mathrm{Br}(X)$ (resp. of $\mathrm{Br}_{nr}(X)$). Note that although we have $X(\mathbb{A}_k)^{\mathrm{Br}(X)} \subset X(k_\Omega)^{\mathrm{Br}_{nr}(X)}$, the induced topology on the first set is in general different from its natural adelic topology. In any case, for a fixed A the function $\sum_{v \in \Omega} \mathrm{inv}_v(A(P_v))$ is locally constant in the adelic or the product topology, respectively. Thus either kernel is a closed subset of its ambient set.

By the Hasse reciprocity law the image of $X(k)$ under the diagonal embedding $X(k) \hookrightarrow X(\mathbb{A}_k)$ is contained in $X(\mathbb{A}_k)^{\mathrm{Br}(X)}$. This implies that the closure of the diagonal image of $X(k)$ in $X(\mathbb{A}_k)$ (resp. in $X(k_\Omega)$) with respect to the adelic (resp. product) topology is contained in $X(\mathbb{A}_k)^{\mathrm{Br}(X)}$ (resp. in $X(k_\Omega)^{\mathrm{Br}_{nr}(X)}$).

A variety X such that $X(\mathbb{A}_k) \neq \emptyset$ whereas $X(k) = \emptyset$ is a *counter-example to the Hasse principle*. Such a counter-example is accounted for by *the Manin obstruction* if $X(\mathbb{A}_k)^{\mathrm{Br}(X)}$ is already empty.

We shall also consider the obstructions defined by various subsets $B \subset \mathrm{Br}(X)$. They are defined by considering the adelic points of X orthogonal to B with respect to (5.1). The obstruction defined by $\mathrm{Br}_1(X)$ will be referred to as *the algebraic Manin obstruction*, as opposed to the obstruction defined by an element of $\mathrm{Br}(X)$ not killed by going over to $\overline{X}$. The obstruction given by such an element is sometimes called 'transcendental'.

The kernel of the Brauer–Manin pairing has the following 'functoriality' property. If $\phi : Y \to X$ is a k-morphism, $\{P_v\} \in Y(\mathbb{A}_k)$, and $A \in \mathrm{Br}(X)$, then

$$\sum_{v \in \Omega} \mathrm{inv}_v((\phi^* A)(P_v)) = \sum_{v \in \Omega} \mathrm{inv}_v(A(\phi(P_v))). \qquad (5.3)$$

In particular, $X(\mathbb{A}_k)^{\mathrm{Br}(X)} = \emptyset$ implies $Y(\mathbb{A}_k)^{\mathrm{Br}(Y)} = \emptyset$. In other words, if Y and X are counter-examples to the Hasse principle, and X is explained by the Manin obstruction, then so is Y. (An example of this situation is a smooth and projective curve C embedded into a k-torsor $Pic^1(C)$ under its Jacobian $Jac(C)$; see Section 6.2.)

If $X(k_\Omega)^{\mathrm{Br}_{nr}(X)}$ is strictly smaller than $X(k_\Omega)$, then since the closure of the diagonal image of $X(k)$ in $X(k_\Omega)$ is contained in the closed subset $X(k_\Omega)^{\mathrm{Br}_{nr}(X)}$, we can conclude that $X(k)$ is not dense in $X(k_\Omega)$. This means that the condition

$$X(k_\Omega) \neq X(k_\Omega)^{\mathrm{Br}_{nr}(X)}$$

is an obstruction to weak approximation on X. It was defined by Colliot-Thélène and Sansuc, and is called *the Brauer–Manin obstruction to weak approximation*. If $X(k)$ is dense in $X(k_\Omega)^{\mathrm{Br}_{nr}(X)}$, we shall say that this obstruction is the only one.

The statement that the Manin obstruction to the Hasse principle and weak approximation is the only obstruction for a class of varieties can be summarized by the formula

$$\overline{X(k)} = X(k_\Omega)^{\mathrm{Br}_{nr}(X)},$$

where $\overline{X(k)}$ is the closure of the diagonal image of $X(k)$. For a proper variety X this formula is the same as

$$\overline{X(k)} = X(\mathbb{A}_k)^{\mathrm{Br}(X)}.$$

For the sake of completeness we extend the above pairings to the study of strong and weak approximation away from Σ.

This can be done when $X(k) \neq \emptyset$. Let $P \in X(k)$. Following Borovoi we consider a variant of the Brauer–Manin pairing:

$$\mathrm{Br}^\Sigma(X) \times X(\mathbb{A}_k^\Sigma) \to \mathbf{Q}/\mathbf{Z}, \quad (A, \{P_v\}) \longmapsto \sum_{v \in \Omega \setminus \Sigma} \mathrm{inv}_v(A(P_v) - A(P)).$$

$$(5.4)$$

It is clear that when $\Sigma = \emptyset$, this is just the adelic Brauer–Manin pairing. It follows from the global reciprocity law that

(1) the sum does not depend on the choice of $P \in X(k)$,
(2) the diagonal image of $X(k)$ belongs to the right kernel of this pairing.

This pairing can be used to define an obstruction to strong approximation away from Σ. (If Σ is a complement of a finite set, then $X(\mathbb{A}_k^\Sigma) = X(k_{\Omega \setminus \Sigma})$, and there is no difference between the strong and the weak approximation at a finite set of places. This is classically referred to as weak approximation at $\Omega \setminus \Sigma$.)

We can also consider an analogue of the product pairing:

$$\mathrm{Br}_{nr}^\Sigma(X) \times X(k^\Sigma) \to \mathbf{Q}/\mathbf{Z}, \quad (A, \{P_v\}) \longmapsto \sum_{v \in \Omega \setminus \Sigma} \mathrm{inv}_v(A(P_v) - A(P)).$$

This pairing produces an obstruction to weak approximation away from Σ.

Let us now list a few typical classes of varieties for which one can prove that the Manin obstruction to the Hasse principle and weak approximation is the only one. We continue to assume cubics and intersections of two quadrics geometrically integral, non-conical and of codimension 1 and 2, respectively.

Theorem 5.2.1 *For the following classes of geometrically integral varieties over a number field k we have* $\overline{X(k)} = X(k_\Omega)^{\mathrm{Br}_{nr}(X)}$:

(a) k-torsors under connected affine algebraic groups [Sansuc 81], *and, more generally, homogeneous spaces of connected affine algebraic groups with connected stabilizers* [Bor 96];

(b) smooth proper models of cubic hypersurfaces with three singular points globally defined over k [CS89], *with two singular points globally defined over k* ([CSS87], [H94], [H95]), *of singular cubic threefolds with isolated singularities* ([CLSS], [SB98]);

(c) smooth proper models of complete intersections of two quadrics in $\mathbf{P}^n_k$, $n \geq 4$, *with two singular points or two lines globally defined over k* [CSS87], *or with a k-point or a k-conic* ([CSS87], [SS], [CS92]).

The list of known results is by no means complete, and is given here only as an illustration. The proofs of some of these results can be found in subsequent chapters.

For k-torsors under abelian varieties A such that $Ш(A)$ is finite, the Manin obstruction to the Hasse principle is the only obstruction [Manin 70]; see Theorem 6.2.3. The same is true for weak approximation provided that rational points are dense in the connected component of the neutral element in $\prod_{v \in \Omega_\infty} A(k_v)$ [Wang].

5.3 Descent obstructions

Let G be an affine algebraic k-group. Recall (cf. Section 2.2) that twisting a right torsor $f : Y \to X$ under G by a cocycle $\sigma \in Z^1(k, G)$ produces a right torsor $f^\sigma : Y^\sigma \to X$ under the twisted group G^σ. This operation commutes with base change; for instance, with taking the fibre Y_P at a k-point of X. In the abelian case, the inner form G^σ can be identified with G and the map $H^1(X, G) \to H^1(X, G^\sigma)$ is just the translation by $-[\sigma]$. Replacing σ by a cohomologous cocycle gives an isomorphic torsor; in particular, the subset $f^\sigma(Y^\sigma(k))$ of $X(k)$ depends only on the class $[\sigma] \in H^1(k, G)$. We shall use the notation $H^1(X, G)$ for the Čech cohomology set $\check{H}^1(X, G)$; this set classifies X-torsors under G up to isomorphism (by (2.10), since G

is affine). We have the following partition of $X(k)$ (cf. (2.12)):

$$X(k) = \bigcup_{[\sigma] \in H^1(k,G)} f^\sigma(Y^\sigma(k)).$$

Descent obstruction to the Hasse principle.

Suppose that $X(\mathbb{A}_k) \neq \emptyset$. 'Evaluating' $f : Y \to X$ at an adelic point of X gives a map

$$\theta_f : X(\mathbb{A}_k) \to \prod_{v \in \Omega} H^1(k_v, G).$$

Note that since G is affine, then the set $H^1(k_v, G)$ is finite ([Serre, CG], III.4). For each $\sigma \in Z^1(k, G)$, we let σ_v denote its image in $Z^1(k_v, G)$. (This image is defined by first choosing a place w of $\overline{k}$ over v, and then restricting σ to the decomposition group D_w of w. The union of completions at w of finite subextensions of $\overline{k}$ is an algebraic closure of k_v, and D_w is its Galois group over k_v; cf. [Serre, CG], p. 115.) The corresponding map of cohomology sets $H^1(k, G) \to H^1(k_v, G)$ sends the class of a torsor T to the class of $T \times_k k_v$.

Definition 5.3.1 *Let X be a smooth and geometrically integral variety over a number field k, Σ be a finite set of places of k. Let $f : Y \to X$ be a torsor under a linear algebraic k-group G. Define $X(\mathbb{A}_k^\Sigma)^f$ as the subset of $X(\mathbb{A}_k^\Sigma)$ consisting of adelic points whose image under the evaluation map $X(\mathbb{A}_k^\Sigma) \to \prod_{v \in \Omega \backslash \Sigma} H^1(k_v, G)$ comes from an element of $H^1(k, G)$, or, in other terms,*

$$X(\mathbb{A}_k^\Sigma)^f = \bigcup_{[\sigma] \in H^1(k,G)} f^\sigma(Y^\sigma(\mathbb{A}_k^\Sigma)).$$

We have $X(k) \subset X(\mathbb{A}_k^\Sigma)^f \subset X(\mathbb{A}_k^\Sigma)$. (When $\Sigma = \emptyset$ we shall omit the sign $\emptyset$ from the notation.) The emptiness of $X(\mathbb{A}_k)^f$ is thus an obstruction to the existence of a k-point on X. That is, the emptiness of $X(\mathbb{A}_k)^f$ when $X(\mathbb{A}_k)$ is non-empty is an obstruction to the Hasse principle. We call it **the descent obstruction** *defined by $f : Y \to X$.*

It is clear from this definition that $X(\mathbb{A}_k)^f$ depends only on the isomorphism class $[Y] \in H^1(X, G)$.

Note that if G is a k-group of multiplicative type, the diagonal image of $H^1(k, G)$ in the product $\prod_{v \in \Omega} H^1(k_v, G)$ is described by the Poitou–Tate exact sequence (cf. [Milne, ADT], I.4.20 (b), I.4.13). There is a generalization of this sequence, due to R. Kottwitz, to the case when G is connected and reductive (cf. [Bor 98], Thm. 5.15).

Proposition 5.3.2 *Let $f : Y \to X$ be a torsor under a linear algebraic group G, and assume that X is a* **proper** *k-variety. Let $\Sigma \subset \Omega$ be a finite set of places. Then there are only finitely many classes $[\sigma] \in H^1(k, G)$ such that $Y^\sigma(k^\Sigma) \neq \emptyset$.*

Proof. Let G^0 be the connected component of G. Then $F = G/G^0$ is a finite k-group.

For a finite set of places $\Sigma' \supset \Sigma$ containing the archimedean ones, let $\mathcal{O}_{k,\Sigma'} \subset k$ be the ring of Σ'-integers of k (integers away from Σ'). Let us fix Σ' large enough so that the following properties hold:

G extends to a smooth group scheme $\mathcal{G}$ over $Spec(\mathcal{O}_{k,\Sigma'})$,

X extends to a proper scheme $\mathcal{X}$ over $Spec(\mathcal{O}_{k,\Sigma'})$,

Y extends to an $\mathcal{X}$-torsor $\mathcal{Y}$ under $\mathcal{G}$.

We denote by $\mathcal{G}^0$ and $\mathcal{F}$ some group schemes over $Spec(\mathcal{O}_{k,\Sigma'})$ extending G^0 and F, respectively. Up to enlarging Σ' we can assume that these group schemes fit into an exact sequence

$$1 \to \mathcal{G}^0 \to \mathcal{G} \to \mathcal{F} \to 1.$$

Let $[\sigma] \in H^1(k, G)$ be such that $Y^\sigma(k^\Sigma) \neq \emptyset$. The condition $Y^\sigma(k_v) \neq \emptyset$ implies that there exists a k_v-point $M_v \in X(k_v)$ such that $[Y_{M_v}] = [\sigma_v]$. By the properness of $\mathcal{X}/\mathcal{O}_{k,\Sigma'}$, for all $v \notin \Sigma'$ we have $X(k_v) = \mathcal{X}(\mathcal{O}_v)$. By our choice of Σ', for all $v \notin \Sigma'$ the class $[\sigma_v]$ is the image of $[\mathcal{Y}_{M_v}]$ under the natural map $H^1(\mathcal{O}_v, \mathcal{G}) \to H^1(k_v, G)$. Thus the image of $[\sigma_v]$ in $H^1(k_v, F)$ comes from $H^1(\mathcal{O}_v, \mathcal{F})$ for all $v \notin \Sigma'$. The image of $[\sigma]$ in $H^1(k, F)$ can be represented by a k-torsor Z under F. This is a 0-dimensional k-scheme, hence $Z = Spec(k[Z])$. The étale k-algebra $k[Z]$ is a product of field extensions of k. The fact that the image of $[\sigma_v]$ in $H^1(k_v, F)$ comes from $H^1(\mathcal{O}_v, \mathcal{F})$ implies that all of these fields are not ramified at all $v \notin \Sigma'$. The degrees of these extensions of k are bounded by $|F(\bar{k})|$. There are only finitely many extensions of k of bounded degree which are unramified away from Σ' ([Lang, ANT], V.4, Thm. 5). In particular, there exists a finite Galois field extension k'/k which contains all these extensions. Thus the image of $[\sigma]$ in $H^1(k, F)$ is contained in a finite subset (the image of $H^1(Gal(k'/k), F)$ in $H^1(k, F)$), which we can take to be the image of a finite subset $\Phi \subset H^1(k, G)$ consisting of elements coming from $H^1(\mathcal{O}_v, \mathcal{G})$ for all $v \notin \Sigma'$. On replacing G with its twist by a cocycle representing a class in Φ, it is now enough to prove that the set of classes $[\sigma] \in H^1(k, G)$ going to 0 in $H^1(k, F)$, and such that for all $v \notin \Sigma'$ we have $[\sigma_v] \in \mathrm{Im}\,[H^1(\mathcal{O}_v, \mathcal{G}) \to H^1(k_v, G)]$, is finite. Let $[\rho_v] \in H^1(\mathcal{O}_v, \mathcal{G})$ be a class mapping to $[\sigma_v]$. We claim that $[\rho_v]$ goes to 0 in $H^1(\mathcal{O}_v, \mathcal{F})$. For this it is enough to show that the kernel of the map of pointed sets $H^1(\mathcal{O}_v, \mathcal{F}) \rightarrow H^1(k_v, F)$ is trivial. To prove this we observe that a $Spec(\mathcal{O}_v)$-torsor under a finite (hence proper) group $\mathcal{F}$ is proper over $Spec(\mathcal{O}_v)$, hence, by the valuative criterion of properness, a section over the generic point $Spec(k_v) \subset Spec(\mathcal{O}_v)$ extends to a section over the whole of $Spec(\mathcal{O}_v)$. Therefore $[\rho_v]$ goes to 0 in $H^1(\mathcal{O}_v, \mathcal{F})$, and hence comes from $H^1(\mathcal{O}_v, \mathcal{G}^0)$. However, every $Spec(\mathcal{O}_v)$-torsor under the smooth and connected group $\mathcal{G}^0$ is trivial by Lang's theorem [Lang 56] (which allows one to find a rational point in the closed fibre) and Hensel's lemma (which allows one to lift it to a section over $Spec(\mathcal{O}_v)$). Thus $H^1(\mathcal{O}_v, \mathcal{G}^0)$ is trivial, hence $[\rho_v] = 1$ implying $[\sigma_v] = 1$ for all $v \notin \Sigma'$. Since every set $H^1(k_v, G)$ is finite, $\{[\sigma_v]\}$ belongs to the finite subset of $\prod_{v \in \Omega \setminus \Sigma} H^1(k_v, G)$ consisting of $\{\alpha_v\}$ such that α_v is arbitrary for $v \in \Sigma' \setminus \Sigma$, and $\alpha_v = 1$ otherwise. By a theorem of Borel and Serre ([Serre, CG], III.4.6; [BS], 7.1) the natural diagonal map $H^1(k, G) \rightarrow \prod_{v \in \Omega \setminus \Sigma} H^1(k_v, G)$ has finite fibres, hence the inverse image of our finite subset is also finite. Thus the set of classes $[\sigma] \in H^1(k, G)$ such that $Y^\sigma(k_v) \neq \emptyset$ for any $v \notin \Sigma$ is finite. QED

Descent obstruction to weak approximation.

For proper varieties the set $X(\mathbb{A}_k^\Sigma)^f$ also provides an obstruction to weak approximation away from Σ. The key fact is that the function associating to a $M_v \in X(k_v)$ the class $[f^{-1}(M_v)] \in H^1(k_v, G)$ is locally constant in the topology of k_v. To see this we can apply twisting, if necessary, and assume that $[f^{-1}(M_v)] = 0$. Then $M_v = f(Q_v)$ for some $Q_v \in Y(k_v)$. By the inverse function theorem for k_v over a small v-adic neighbourhood of M_v we can find a section of f passing through Q_v. Thus the class of the fibre is also 0 for all k_v-points in this neighbourhood.

Proposition 5.3.3 *Let X be a* **proper***, smooth and geometrically integral variety such that $X(k) \neq \emptyset$. Let $\overline{X(k)}^\Sigma$ be the closure of the image of $X(k)$ in $X(\mathbb{A}_k^\Sigma)$; then*

$$\overline{X(k)}^\Sigma \subset X(\mathbb{A}_k^\Sigma)^f.$$

Proof. Using Proposition 5.3.2, one can find a finite set $\Lambda \subset H^1(k, G)$ such that $Y^\sigma(\mathbb{A}_k^\Sigma) = \emptyset$ for $[\sigma] \notin \Lambda$. Therefore, the union in Definition 5.3.1 is actually finite. Now it is enough to show that $f(Y(\mathbb{A}_k^\Sigma))$ is closed in $X(\mathbb{A}_k^\Sigma)$.

Let $\{M_v\}$ be in the closure of $f(Y(\mathbb{A}_k^\Sigma))$. For any $v \notin \Sigma$ let U_v be a small neighbourhood of $M_v \in X(k_v)$ in the corresponding local topology such that $[f^{-1}(M'_v)] = [f^{-1}(M_v)] \in H^1(k_v, G)$ for any $M'_v \in U_v$. The open set U_v contains the image $f(Q_v)$ of a local point $Q_v \in Y(k_v)$. Therefore $[f^{-1}(M_v)] = [f^{-1}(f(Q_v))] = 0$, which means that $M_v = f(P_v)$ for some $P_v \in Y(k_v)$. Hence $\{M_v\} \in f(Y(\mathbb{A}_k^\Sigma))$ which proves that $f(Y(\mathbb{A}_k^\Sigma))$ is closed. QED

This shows that the condition $X(\mathbb{A}_k)^f \neq X(\mathbb{A}_k)$ (resp. $X(\mathbb{A}_k^\Sigma)^f \neq X(\mathbb{A}_k^\Sigma)$) is an obstruction to weak approximation (resp. to weak approximation outside Σ) on X.

We shall call this condition *the descent obstruction* to weak approximation (resp. to weak approximation outside Σ) associated to $f : Y \to X$. Note that unlike the Brauer–Manin condition it is only defined for proper varieties X.

One says that the descent obstruction to the Hasse principle and weak approximation related to the torsor $f : Y \to X$ is the only one if

$$\overline{X(k)} = X(\mathbb{A}_k)^f.$$

The Manin obstruction as a particular case of the descent obstruction.

Conditionally on a widely believed conjecture one can realize the Manin obstruction given by an element of $Br(X) = H^2(X, \mathbf{G}_m)$ by the descent obstruction related to a certain X-torsor under PGL_n.

Let X be smooth. We let $\mathrm{Br}_A(X)$ denote the Brauer group of X, defined as the group of similarity classes of Azumaya algebras over X. A theorem of Grothendieck states that there is an injection $\mathrm{Br}_A(X) \to \mathrm{Br}(X)$ ([Milne, EC], IV.2.5). More precisely, the exact sequence of étale sheaves

$$1 \to \mathbf{G}_m \to GL_n \to PGL_n \to 1$$

gives rise to the exact sequence of pointed (Čech cohomology) sets

$$H^1(X, \mathbf{G}_m) \to H^1(X, GL_n) \to H^1(X, PGL_n) \xrightarrow{d_n} \mathrm{Br}(X).$$

The group $\mathrm{Br}_A(X) \subset \mathrm{Br}(X)$ is the union of images of $d_n(H^1(X, PGL_n))$ for all n. It is conjectured that in fact $\mathrm{Br}_A(X) = \mathrm{Br}(X)$. (Some partial results are obtained by O. Gabber and R. Hoobler.) It is known that $d_n(H^1(X, PGL_n)) \subset \mathrm{Br}_A(X)[n]$ ([Milne, EC], IV.2.7). In the case $X = Spec(k)$ where k is a number field or a local field, it is well known that the map

$$d_n : H^1(k, PGL_n) \to \mathrm{Br}(k)[n]$$

is surjective. (The order of the class of a central simple algebra in the

Brauer group of k equals its index.) This map is also injective (see [Serre, CL], X.5), and hence is bijective.

Let **PGL** be the disjoint union of sets $H^1(X, PGL_n)$ for all $n = 2, 3, \ldots$.

Proposition 5.3.4 *We have*

$$X(\mathbb{A}_k)^{\mathrm{Br}_A(X)} = \bigcap_{f \in \mathbf{PGL}} X(\mathbb{A}_k)^f.$$

Proof. Let $\alpha \in \mathrm{Br}_A(X)$. Then there exist an integer n and a torsor $f : Y \to X$ under PGL_n such that $\alpha = d_n([Y])$. Then $n\alpha = 0$.

Let $\{M_v\} \in X(\mathbb{A}_k)$. The following diagram, where the upper vertical maps are specializations at $\{M_v\}$, and the lower ones are the natural diagonal maps, is commutative:

$$
\begin{array}{ccc}
H^1(X, PGL_n) & \xrightarrow{d_n} & \mathrm{Br}(X)[n] \\
\downarrow & & \downarrow \\
\prod_{v \in \Omega} H^1(k_v, PGL_n) & \xrightarrow{d_n} & \prod_{v \in \Omega} \mathrm{Br}(k_v)[n] \\
\uparrow & & \uparrow \\
H^1(k, PGL_n) & \xrightarrow{d_n} & \mathrm{Br}(k)[n]
\end{array}
$$

The commutativity of this diagram and the bijectivity of its middle and bottom horizontal maps imply that

$$\{[Y_{M_v}]\} \in Im\big[H^1(k, PGL_n) \to \prod_{v \in \Omega} H^1(k_v, PGL_n)\big]$$

if and only if

$$\{\alpha(M_v)\} \in Im\big[\mathrm{Br}(k) \to \prod_{v \in \Omega} \mathrm{Br}(k_v)\big].$$

In other words, $X(\mathbb{A}_k)^f = X(\mathbb{A}_k)^\alpha$. Since $\mathrm{Br}_A(X)$ is the union of the images of $H^1(X, PGL_n)$ in $\mathrm{Br}(X)$ for $n = 1, 2, \ldots$, we are done. QED

Exercise

Using Hasse's norm theorem prove a theorem of Selmer that (non-conical) diagonal cubic surfaces given by the equation $ax^3 + by^3 + cz^3 + dt^3 = 0$ with $ab = cd$ satisfy the Hasse principle.

Comments

After Minkowski and Hasse the study of the Hasse principle followed two natural routes: (A) from quadrics to more general homogeneous spaces of algebraic groups, and (B) from quadratic equations to Diophantine equations of higher degree. To the list of positive results mentioned above, let us add the following. The study of the Hasse principle through descent led Selmer (1951) to the discovery of the curve $3x^3 + 4y^3 + 5z^3 = 0$ which is a counter-example to the Hasse principle. This was later put into a conceptual framework with the introduction of the Tate–Shafarevich group. It was conjectured by Mordell (1949) that non-conical cubic surfaces satisfy the Hasse principle. The first counter-example was found by Swinnerton-Dyer in 1962. Selmer, motivated by his theorem that diagonal cubic surfaces $ax^3 + by^3 + cz^3 + dt^3 = 0$ such that $ab = cd$ satisfy the Hasse principle, tried to rescue Mordell's conjecture by restricting it to diagonal cubic surfaces. This also turned out to be false, as revealed by the counter-example of Cassels and Guy quoted in the Introduction. In [Manin, CF] Manin showed that the obstruction now bearing his name explains all counter-examples to the Hasse principle known by that time (it was less clear in the case of Cassels and Guy). Colliot-Thélène, Kanevsky and Sansuc ([CKS], [C86]) computed the Manin obstruction for diagonal cubic surfaces over $\mathbf{Q}$. A computer search for integral coefficients of absolute value up to 100 showed that whenever the Manin obstruction disappears the diagonal cubic surface has a rational point. In particular, this settled the case of the counter-example of Cassels and Guy. Colliot-Thélène and Sansuc conjectured that the absence of the Manin obstruction is a necessary and sufficient condition for the Hasse principle for all smooth and proper rational surfaces. Recently, a new (conditional) case of the Hasse principle for diagonal cubic surfaces over $\mathbf{Q}$ was proved by Heath-Brown, see Theorem 6.2.7. Swinnerton-Dyer then went further and combined the method of Heath-Brown with his own powerful technique and proved a much stronger Theorem 6.2.8.

In the case of smooth and proper rational varieties the geometric Brauer group $\mathrm{Br}(\overline{X})$ is trivial. One can ask whether an element of $\mathrm{Br}(X)$ which survives in $\mathrm{Br}(\overline{X})$ can give a non-trivial obstruction. D. Harari [H96] has constructed (by explicit equations) a smooth and projective threefold X over $k = \mathbf{Q}$ with $X(\mathbb{A}_k) \neq \emptyset$, $X(\mathbb{A}_k)^{\mathrm{Br}_A(X)} = \emptyset$ and $\mathrm{Br}_1(X) = \mathrm{Br}_0(X)$. The geometric Brauer group is $\mathrm{Br}(\overline{X}) = \mathbf{Z}/2$. Hence this non-trivial Manin obstruction is given by an Azumaya algebra which survives in $\mathrm{Br}(\overline{X})$. It follows from Theorem 6.1.1 below that such a 'transcendental' Manin obstruction cannot be given by a descent obstruction associated to a torsor under an abelian algebraic group. Fortunately, by Proposition 5.3.4, it can still be given by a non-abelian descent obstruction. This is at least one justification for introducing the non-abelian descent obstructions.

For the sake of completeness let us note that in [HS] there is also defined another obstruction to the existence of rational points on a variety over an arbitrary field k. It is the obstruction for a 'descent datum' on a torsor $\overline{f} : \overline{Y} \to \overline{X}$ defined over $\overline{k}$ to come from a torsor $f : Y \to X$ defined over k. In the abelian case this reduces to the elementary obstruction of Colliot-Thélène and Sansuc (cf. Proposition 2.2.4 and Section 2.3). In the non-abelian case one has to use the machinery of second non-abelian cohomology. In this book we consider only two particular cases: the case when f is finite and étale, and the case when X is a homogeneous space of an algebraic group (see Section 2.4).

6

Abelian descent and Manin obstruction

In this chapter we begin to collect the fruits of our long journey. The first section is devoted to the statement and the proof of the main theorem of the descent theory of Colliot-Thélène and Sansuc. The original formulation of Colliot-Thélène and Sansuc is particularly clear: when $Pic(\overline{X})$ is torsion free, then an adelic point on X orthogonal to $\mathrm{Br}_1(X)$ is the image of an adelic point on a universal torsor under the structure map. The converse is also true, namely, the image of an adelic point on a universal torsor is orthogonal to $\mathrm{Br}_1(X)$. This theorem reduces the proof of the fact that the Manin obstruction to the Hasse principle and weak approximation is the only one to the analysis of these properties for torsors. This method will be illustrated in the next chapter on the example of conic bundles.

In the second section we prove that the Manin obstruction to the Hasse principle is the only obstruction for principal homogeneous spaces of certain classes of algebraic groups: tori, semisimple groups, abelian varieties (conditional on the finiteness of the Tate–Shafarevich group). Also discussed are some other, more or less related results of this kind. We look at the Manin obstruction on curves and discuss what is known here. We then state the theorems of Heath-Brown and Swinnerton-Dyer on the existence of rational points on diagonal cubic (hyper)surfaces. All these results are based on the global duality for abelian varieties; this is our reason for including them here.

In the final section we prove that the Manin obstruction to the Hasse principle and weak approximation is the only obstruction for smooth compactifications of principal homogeneous spaces of tori, and prove other results needed in the next chapter.

112

6.1 Descent theory

The main result of this section is the following theorem.

Theorem 6.1.1 *Let X be a variety over a number field k such that $\overline{k}[X]^* = \overline{k}^*$; then we have*

$$X(\mathbb{A}_k)^{\mathrm{Br}_1(X)} = \bigcap_{\lambda:\, M \hookrightarrow Pic(\overline{X})} \; \bigcup_{\mathrm{type}(Y,f)=\lambda} f(Y(\mathbb{A}_k)),$$

where $\lambda : M \hookrightarrow Pic(\overline{X})$ runs over the Γ_k-submodules of $Pic(\overline{X})$ of finite type.

This means that the algebraic Manin obstruction is equivalent to the combination of obstructions of two different kinds: the obstruction for the existence of torsors of a given type λ, and the descent obstruction defined by torsors of type λ, for all possible λ's.

This theorem is a consequence of the following more detailed result.

Let $r : \mathrm{Br}_1(X) \to H^1(k, Pic(\overline{X}))$ be the canonical map from the Hochschild–Serre spectral sequence $H^p(k, H^q(\overline{X}, \mathbf{G}_m)) \Rightarrow H^{p+q}(X, \mathbf{G}_m)$ (see (2.23)). Let M be a Γ_k-module of finite type, and $\lambda : M \to Pic(\overline{X})$ a homomorphism of Γ_k-modules. Let S be the k-group of multiplicative type such that $M = \hat{S}$. Recall that

$$\mathrm{Br}_\lambda(X) := r^{-1}\lambda_*(H^1(k, M)) \subset \mathrm{Br}_1(X).$$

We define $X(\mathbb{A}_k)^{\mathrm{Br}_\lambda} \subset X(\mathbb{A}_k)$ as the set of adelic points orthogonal to $\mathrm{Br}_\lambda(X)$ with respect to the adelic Brauer–Manin pairing.

Theorem 6.1.2 *Let X be a variety over a number field k such that $\overline{k}[X]^* = \overline{k}^*$, M be a Γ_k-module of finite type, S its dual group of multiplicative type, and $\lambda \in Hom_k(M, Pic(\overline{X}))$. Then*

(a) we have

$$X(\mathbb{A}_k)^{\mathrm{Br}_\lambda} = \bigcup_{\mathrm{type}(Y,f)=\lambda} f(Y(\mathbb{A}_k)), \tag{6.1}$$

where $f : Y \to X$ runs over the set of X-torsors under S of type λ,

(b) if X is proper, there are only finitely many isomorphism classes of torsors $f : Y \to X$ of type λ such that $Y(k_\Omega) \neq \emptyset$ (and hence also such that $Y(\mathbb{A}_k) \neq \emptyset$).

When X is proper we can also write

$$X(k_\Omega)^{\mathrm{Br}_\lambda} = \bigcup_{\mathrm{type}(Y,f)=\lambda} f(Y(k_\Omega)). \tag{6.2}$$

Part *(b)* follows from Proposition 5.3.2 and the fact that X-torsors of a given type λ can be obtained from one another by twisting by a 1-cocycle with coefficients in S (see Section 2.3). Formula (6.2) follows from (6.1) since $X(\mathbb{A}_k) = X(k_\Omega)$ when X is proper. (To conclude that the right hand side of (6.2) is contained in the left hand side use the same argument as in the proof of (3) at the end of the proof of Theorem 6.1.2 *(a)*.)

To derive Theorem 6.1.1 note that for any $\alpha \in \mathrm{Br}_1(X)$ there exists a Γ_k-submodule $\lambda : M \hookrightarrow Pic(\overline{X})$ of finite type such that $r(\alpha) \in \lambda_*(H^1(k, M))$ ([Serre, CG], I.2.2, Cor. 2). Thus $X(\mathbb{A}_k)^{\mathrm{Br}_1(X)} = \bigcap_\lambda X(\mathbb{A}_k)^{\mathrm{Br}_\lambda}$.

Let us point out some of the many corollaries of this theorem (keeping the assumption $\overline{k}[X]^* = \overline{k}^*$):

Corollary 6.1.3 *(1) The Manin obstruction to the Hasse principle related to $\mathrm{Br}_\lambda(X)$ is empty if and only if there exists an X-torsor Y of type λ such that $Y(\mathbb{A}_k) \neq \emptyset$. In particular, when $Pic(\overline{X})$ is of finite type, the vanishing of the algebraic Manin obstruction is equivalent to the existence of universal torsors with an adelic point.*

(2) If the X-torsors of type λ satisfy the Hasse principle, then the Manin obstruction to the Hasse principle on X related to $\mathrm{Br}_\lambda(X)$ is the only obstruction.

(3) Let X be proper. If the X-torsors of type λ satisfy weak approximation, then the Brauer–Manin obstruction to weak approximation on X related to $\mathrm{Br}_\lambda(X)$ is the only obstruction.

Example.

Let X be a smooth proper curve of genus 1 over a field k of characteristic 0, $E = Jac(X)$. Define λ_n as the composite map $E[n](\overline{k}) \to E(\overline{k}) \to Pic(\overline{X})$. Then an X-torsor of type λ_n exists if and only if the class $[X] \in H^1(k, E)$ is divisible by n, and the X-torsors of this type are the k-torsors Y under E such that $n[Y] = [X]$ (Proposition 3.3.5). Thus when k is a number field, by Theorem 6.1.2 *(a)* $X(\mathbb{A}_k)^{\mathrm{Br}_{\lambda_n}} \neq \emptyset$ if and only if the class $[X] \in \text{Ш}(E)$ is divisible by n in $\text{Ш}(E)$. Every element of $H^1(k, E)$ has finite order, hence $\mathrm{Br}(X)$ is the union of $\mathrm{Br}_{\lambda_n}(X)$ for all n. Therefore $X(\mathbb{A}_k)^{\mathrm{Br}} \neq \emptyset$ if and only if $[X]$ belongs to the divisible subgroup of $\text{Ш}(E)$. One actually proves a more precise statement: $[X]$ belongs to the divisible subgroup of $\text{Ш}(E)$ if and only if $X(\mathbb{A}_k)^{\mathrm{B}(X)} \neq \emptyset$ (cf. Theorem 6.2.3).

We shall need the fundamental fact that $H^3(k, \overline{k}^*) = 0$ when k is a local or global field ([CF], VII, 11.4).

Proof of Theorem 6.1.2 (a). We break *(a)* into three statements:

(1) If there exists an adelic point which is Brauer–Manin orthogonal to $r^{-1}(\lambda_*(\text{III}^1(k, M))) \subset \text{Br}_\lambda(X)$, then there exists a torsor $f : Y \to X$ of type λ.

(2) Suppose there exists a torsor $f : Y \to X$ of type λ. If an adelic point is Brauer–Manin orthogonal to $\text{Br}_\lambda(X)$, then there exists $\sigma \in H^1(k, S)$ such that this point lifts to an adelic point on Y^σ.

(3) If there exists a torsor $f : Y \to X$ of type λ such that $Y(\mathbb{A}_k) \neq \emptyset$, then $f(Y(\mathbb{A}_k)) \subset X(\mathbb{A}_k)^{\text{Br}_\lambda}$.

The proof of (1) given below is rather long, so let us first explain its main idea. Recall that we have canonical isomorphisms

$$\epsilon : Ext^i_k(M, \overline{k}^*) = H^i(k, S);$$

see Lemma 2.3.7 or [Milne, ADT], I.0.8.

It follows from the theory of torsors under groups of multiplicative type developed in Section 2.3 that there exists an X-torsor under S of type λ if and only if the image $\partial(\lambda)$ of λ in $H^2(k, S) = Ext^2_k(M, \overline{k}^*)$ is 0. Moreover, $\partial(\lambda) = \lambda^*(e(X))$, where $e(X) \in Ext^2_k(Pic(\overline{X}), \overline{k}^*)$ is the inverse of the class of the following 2-fold extension:

$$1 \to \overline{k}^* \to \overline{k}(X)^* \to Div(\overline{X}) \to Pic(\overline{X}) \to 0. \tag{6.3}$$

We show that $e(X)$ can be interpreted as an element

$$b_X \in \text{III}^2(k, Hom(Pic(\overline{X}), \overline{k}^*)).$$

For any Γ_k-module M one defines the global Poitou–Tate pairing ([Milne, ADT], I.4.20 (a)):

$$\langle \cdot, \cdot \rangle : \text{III}^2(k, Hom(M, \overline{k}^*)) \times \text{III}^1(k, M) \to \mathbf{Q}/\mathbf{Z}.$$

This definition makes sense for any Γ_k-module M, although the Poitou–Tate duality theorem (the non-degeneracy of this pairing) applies only to M of finite type.

We shall prove the following fundamental property: for any adelic point $\{P_v\}$ and any $\alpha \in \text{III}^1(k, M)$ we have

$$\langle b_X, \lambda_*(\alpha) \rangle = \sum_{v \in \Omega} \text{inv}_v(A(P_v)), \tag{6.4}$$

where $A \in \mathcal{B}(X)$ is such that $r(A) = \lambda_*(\alpha)$. By the functoriality of the global Poitou–Tate pairing and the absence of the Manin obstruction related to $Б(X)$ we then obtain

$$\langle \epsilon(\partial(\lambda)), \alpha \rangle = \langle \epsilon(\lambda^*(e(X))), \alpha \rangle = 0 \tag{6.5}$$

for any $\alpha \in \text{III}^1(k, M)$. By the non-degeneracy of the Poitou–Tate pairing we conclude that $\partial(\lambda) = 0$. Hence there exist X-torsors of type λ. This will be enough to prove (1).

For future reference we formulate the following consequence of (6.4) and the Poitou–Tate duality theorem.

Proposition 6.1.4 *Let X be a geometrically integral smooth variety over a number field k such that $\overline{k}[X]^* = \overline{k}^*$ and $Pic(\overline{X})$ is of finite type. Then the universal X-torsors exist (equivalently, $e(X) = 0$) if and only if the Manin obstruction to the Hasse principle on X associated to $Б(X)$ vanishes.*

Proof of (1). This will consist of several steps. In Steps 0 to 2 the field k is any field of characteristic 0.

In the proof $C^i(\Gamma_k, M)$ is the group of continuous i-cochains of Γ_k with coefficients in a discrete Γ_k-module M, d is the differential $C^i(\Gamma_k, M) \to C^{i+1}(\Gamma_k, M)$, $Z^i(\Gamma_k, M) = Ker(d)$ is the group of i-cocycles, $B^i(\Gamma_k, M) = Im(d)$ is the group of i-coboundaries, $H^i(k, M) = Z^i(\Gamma_k, M)/B^i(\Gamma_k, M)$.

Step 0. Representation of the map $r : \text{Br}_1(X) \to H^1(k, Pic(\overline{X}))$. Let $A \in Б(X) \subset \text{Br}_1(X)$. Then $r(A) \in \text{III}^1(k, Pic(\overline{X}))$.

Recall the following commutative diagram (4.17) from Section 4.2:

$$
\begin{array}{ccccc}
\text{Br}(k) & \to & \text{Br}_1(X) & \to & H^1(k, Pic(\overline{X})) \\
\downarrow & & \downarrow & & \downarrow \\
H^2(k, \overline{k}(X)^*) & = & H^2(k, \overline{k}(X)^*) & \to & 0 \\
\downarrow & & \downarrow & & \\
\end{array}
$$
$$
H^1(k, Pic(\overline{X})) \to H^2(k, \overline{k}(X)^*/\overline{k}^*) \to H^2(k, Div(\overline{X}))
$$

$$\tag{6.6}$$

Let $A \in \text{Br}_1(X)$. We can consider the value A_η of A at the generic point $\eta = Spec(k(X))$ of X. Then A_η is split over $\overline{k}(X)$, hence its class in $\text{Br}(k(X))$ is given by a 2-cocycle $f = (f_{s,t})$ in $Z^2(\Gamma_k, \overline{k}(X)^*)$. As the columns of (6.6) are exact, the class of f goes to 0 under the map

$$div_* : H^2(k, \overline{k}(X)^*) \to H^2(k, Div(\overline{X})).$$

Thus $div(f_{s,t})$ is a coboundary ${}^t(D_s) - D_s$ for some divisors $D_s \in Div(\overline{X})$:

$$div(f) = dD, \text{ where } D = (D_s) \in C^1(\Gamma_k, Div(\overline{X})).$$

The map div_* is the composition

$$H^2(k, \overline{k}(X)^*) \to H^2(k, \overline{k}(X)^*/\overline{k}^*) \to H^2(k, Div(\overline{X})).$$

Hence the image of the cohomology class of f in $H^2(k, \overline{k}(X)^*/\overline{k}^*)$ comes from the cohomology class of $cl(D) = (cl(D_s)) \in Z^1(\Gamma_k, Pic(\overline{X}))$ under the boundary map of the exact sequence of Γ_k-modules

$$1 \to \overline{k}(X)^*/\overline{k}^* = Div_0(\overline{X}) \to Div(\overline{X}) \to Pic(\overline{X}) \to 0.$$

Note that $Div(\overline{X})$ is a permutation Γ_k-module, hence $H^1(k, Div(\overline{X})) = 0$ by Shapiro's lemma. Therefore, the class obtained in $H^1(k, Pic(\overline{X}))$ is uniquely defined. We claim that this class is $-r(A) \in H^1(k, Pic(\overline{X}))$. This follows from Lemma 4.3.2 with $i = 2$ when the diagram (4.27) is the diagram of complexes of abelian groups obtained from (4.14) by applying the functor $\mathbf{H}(k, \cdot)$ (the derived functor of $M \mapsto M^{\Gamma_k}$), and U is replaced by the generic point $Spec(k(X))$.

Step 1. Consider the spectral sequence

$$H^p(k, Ext^q(Pic(\overline{X}), \overline{k}^*)) \Rightarrow Ext_k^{p+q}(Pic(\overline{X}), \overline{k}^*).$$

Here Ext^i is the Ext in the category of abelian groups. The abelian group $\overline{k}^*$ is divisible, hence it is an injective object in the category of abelian groups. Therefore we have $Ext^i(\cdot, \overline{k}^*) = 0$ for $i > 0$, the spectral sequence degenerates, and we get

$$Ext_k^2(Pic(\overline{X}), \overline{k}^*) = H^2(k, Hom(Pic(\overline{X}), \overline{k}^*)).$$

Let $b_X \in H^2(k, Hom(Pic(\overline{X}), \overline{k}^*))$ be the image of $e(X)$.

Step 2. Representation of b_X. Let us now choose a 2-cocycle representing the cohomology class b_X.

Choose $\overline{P} \in X(\overline{k})$. Let $\mathcal{O}_{\overline{X}, \overline{P}}$ be the local ring of $\overline{X}$ at $\overline{P}$. Then the natural injection of abelian groups $\overline{k}^* \hookrightarrow \mathcal{O}^*_{\overline{X}, \overline{P}}$ has a section given by $g \mapsto g(\overline{P})$. On the other hand, the natural extension of abelian groups

$$1 \to \mathcal{O}^*_{\overline{X}, \overline{P}} \to \overline{k}(X)^* \to Div_{\overline{P}}(\overline{X}) \to 0,$$

where $Div_{\overline{P}}(\overline{X}) \subset Div(\overline{X})$ consists of divisors passing through $\overline{P}$, is split because $Div_{\overline{P}}(\overline{X})$ is torsion free. Hence there is a homomorphic section $e_{\overline{P}}$ of the natural injection $\overline{k}^* \hookrightarrow \overline{k}(X)^*$ given by $g \mapsto g(\overline{P})$ provided g is invertible at $\overline{P}$. Note that when $\overline{P} = P \in X(k)$ the section e_P can be made Galois equivariant by Proposition 2.3.4 *(b)*.

Let $\sigma_{\overline{P}} : Div_0(\overline{X}) \to \overline{k}(X)^*$ be given by $\sigma_{\overline{P}}(div(g)) = g/e_{\overline{P}}(g)$. (This definition does not depend on the choice of the function g with divisor

$div(g)$, because such a g is uniquely defined up to a constant.) Then $\sigma_{\overline{P}}$ is a homomorphic section of the surjective map in the exact sequence

$$1 \to \overline{k}^* \to \overline{k}(X)^* \to Div_0(\overline{X}) \to 0,$$

such that $\sigma_{\overline{P}}(div(g)) = g/g(\overline{P})$ if g is invertible at $\overline{P}$. Note that if $\overline{P}'$ is another $\overline{k}$-point of X, and the function g is also invertible at $\overline{P}'$, then $\sigma_{\overline{P}}(div(g))/\sigma_{\overline{P}'}(div(g)) \in \overline{k}^*$ equals $g(\overline{P}')/g(\overline{P})$.

The coboundary of $\sigma_{\overline{P}} \in C^0(\Gamma_k, Hom(Div_0(\overline{X}), \overline{k}(X)^*))$ can be viewed as a 1-cocycle $d\sigma_{\overline{P}} \in Z^1(\Gamma_k, Hom(Div_0(\overline{X}), \overline{k}^*))$. Note that (in the additive notation) we have $\sigma_{\overline{P}}(div(\cdot)) + e_{\overline{P}}(\cdot) = 1$, hence $d\sigma_{\overline{P}}$ is the inverse of $de_{\overline{P}}$ (which equals $g(\overline{P})/g(s\overline{P})$ whenever this value is in $\overline{k}^*$).

By the injectivity of $\overline{k}^*$, $d\sigma_{\overline{P}}$ extends to $\psi_{\overline{P}} \in C^1(\Gamma_k, Hom(Div(\overline{X}), \overline{k}^*))$. Finally, we consider the coboundary $d\psi_{\overline{P}} \in B^2(\Gamma_k, Hom(Div(\overline{X}), \overline{k}^*))$. Its restriction to $Div_0(\overline{X})$ is $d^2\sigma_{\overline{P}} = 0$, hence $d\psi_{\overline{P}} \in Z^2(\Gamma_k, Hom(Pic(\overline{X}), \overline{k}^*))$. This 2-cocycle represents the class of (6.3), that is, the cohomology class $-b_X \in H^2(k, Hom(Pic(\overline{X}), \overline{k}^*))$.

Step 3. Let us recall the construction of the global Poitou–Tate pairing

$$\langle \cdot, \cdot \rangle : \text{III}^2(k, Hom(M, \overline{k}^*)) \times \text{III}^1(k, M) \to \mathbf{Q}/\mathbf{Z}.$$

Let $\beta \in \text{III}^2(k, Hom(M, \overline{k}^*))$ and $\alpha \in \text{III}^1(k, M)$ be represented by cocycles b and a. Then $b \cup a = dh$, where $h \in C^2(\Gamma_k, \overline{k}^*)$, because $H^3(\Gamma_k, \overline{k}^*) = 0$ for any number field k. For any place v the restriction of b to the decomposition group $\Gamma_v \subset \Gamma_k$ has the form $d\xi_v$, where $\xi_v \in C^1(\Gamma_v, Hom(M, \overline{k}_v^*))$. Then $\xi_v \cup a - h$ is a 2-cocycle. Let $\epsilon_v \in Br(k_v)$ be its class. The Poitou–Tate pairing is defined as

$$\langle \beta, \alpha \rangle = \sum_{v \in \Omega} inv_v(\epsilon_v) \in \mathbf{Q}/\mathbf{Z}.$$

Since $X(\mathbb{A}_k) \neq \emptyset$, $e(X)$ goes to 0 under the restrictions from k to k_v for all places v (Proposition 2.3.4 *(b)*), so that

$$b_X \in \text{III}^2(k, Hom(Pic(\overline{X}), \overline{k}^*)).$$

Let us consider the global Poitou–Tate pairing for $M = Pic(\overline{X})$. We begin our computation of $\langle b_X, r(A) \rangle = \langle b_X, \lambda_*(\alpha) \rangle = \langle d\psi_{\overline{P}}, cl(D) \rangle$ (note that two minus signs get cancelled) by constructing the cochain h (the global ingredient of the pairing). Define $h \in C^2(\Gamma_k, \overline{k}^*)$ by $h = \psi_{\overline{P}} \cup D - e_{\overline{P}} \cup f$, where $f = (f_{s,t})$ and $D = (D_s)$ are from Step 0. Then $dh = d\psi_{\overline{P}} \cup cl(D)$. Indeed, we have $d(f) = 0$, hence

$$dh = d\psi_{\overline{P}} \cup D - \psi_{\overline{P}} \cup dD - de_{\overline{P}} \cup f = d\psi_{\overline{P}} \cup D,$$

because $\psi_{\overline{P}}$ applied to the class of f in $Z^2(\Gamma_k, \overline{k}(X)^*/\overline{k}^*)$ is just $d\sigma_{\overline{P}} \cup f$.

Step 4. Let k be a number field, and v be a place of k. Now we construct the local factor of the pairing at v.

We choose $P_v \in X(k_v)$ away from $div(f_{s,t})$, $s, t \in \Gamma_k$. This is always possible by the implicit function theorem. Let $\overline{k}_v$ be an algebraic closure of k_v; the Galois group $Gal(\overline{k}_v/k_v)$ is isomorphic to the decomposition group Γ_v. As was explained in Step 2 we can choose σ_{P_v} to be Γ_v-equivariant. Define a homomorphism $\theta_v : Div_0(\overline{X}) \to \overline{k}_v^*$ as the composition of the natural inclusion $Div_0(\overline{X}) \hookrightarrow Div_0(\overline{X}_v)$ with $\sigma_{\overline{P}}/\sigma_{P_v} : Div_0(\overline{X}_v) \to \overline{k}_v^*$. It has the property that $\theta_v(div(g)) = g(P_v)/g(\overline{P})$ whenever a rational function $g \in \overline{k}(X)^*$ is invertible at $\overline{P}$ and P_v (see Step 2). Since σ_{P_v} is Γ_v-equivariant, we have $d\sigma_{P_v} = 0$, and therefore $d\theta_v$ is the restriction of $d\sigma_{\overline{P}} \in Z^1(\Gamma_k, Hom(Div_0(\overline{X}), \overline{k}^*))$ to k_v (this restriction was described at the beginning of Section 5.3).

Since $\overline{k}_v^*$ is injective as an abelian group, θ_v extends to a homomorphism $\mu_v : Div(\overline{X}) \to \overline{k}_v^*$. Define

$$\xi_v = \psi_{\overline{P}} - d\mu_v \in C^1(\Gamma_v, Hom(Div(\overline{X}), \overline{k}_v^*)).$$

In fact, ξ_v is trivial on $Div_0(\overline{X})$ because the restriction of $\psi_{\overline{P}}$ to $Div_0(\overline{X})$ is $d\sigma_{\overline{P}}$ by the construction of $\psi_{\overline{P}}$, and the restriction of $d\mu_v$ is $d\theta_v = res_{k,k_v}(d\sigma_{\overline{P}})$. Hence

$$\xi_v \in C^1(\Gamma_v, Hom(Pic(\overline{X}), \overline{k}_v^*)).$$

Moreover, $d\xi_v$ is just $res_{k,k_v}(d\psi_{\overline{P}})$. Since $d\psi_{\overline{P}}$ represents the cohomology class $-b_X \in H^2(k, Hom(Pic(\overline{X}), \overline{k}^*))$, ξ_v is the required local factor.

Step 5. Following Step 3 we define $\epsilon_v \in \mathrm{Br}(k_v)$ as the class of the cocycle

$$\begin{aligned} \xi_v \cup \mathrm{cl}(D) - h &= \psi_{\overline{P}} \cup D - d\mu_v \cup D - \psi_{\overline{P}} \cup D + e_{\overline{P}} \cup f \\ &= -d(\mu_v \cup D) + \mu_v \cup dD + e_{\overline{P}} \cup f. \end{aligned}$$

Since $\mu_v \cup D \in C^1(\Gamma_v, \overline{k}_v^*)$, the element $\epsilon_v \in \mathrm{Br}(k_v)$ can also be represented by $\theta_v \cup div(f) + e_{\overline{P}} \cup f$. The first summand equals $f(P_v)/f(\overline{P})$, and the second one $f(\overline{P})$. Therefore, $\epsilon_v \in \mathrm{Br}(k_v)$ is the class of the 2-cocycle $f(P_v) = (f_{s,t}(P_v))$. This proves that $\mathrm{inv}_v(\xi_v \cup D - h) = \mathrm{inv}_v(A(P_v))$. Formula (6.4) is established. This finishes the proof of (1).

Proof of (2). Let $\mathcal{O}_k \subset k$ be the ring of integers of k. Let T be a finite set of places of k containing all the archimedean places. T can be enlarged so that the ring $\mathcal{O}_{k,T} \subset \mathcal{O}_k$ obtained by inverting the primes corresponding to finite places of T is such that M is unramified over $\mathcal{O}_{k,T}$. Let $S =$

$Hom_{\mathcal{O}_{k,T}\text{-groups}}(M, \mathbf{G}_m)$. Let $\mathcal{O}_v$ be the completion of $\mathcal{O}_k$ at a prime v, $S_v := S \times_{\mathcal{O}_T} \mathcal{O}_v$ for $v \notin T$. To simplify notation we shall write $H^i(R, \cdot)$ for $H^i(Spec(R), \cdot)$ when R is a ring. We write $N^* = Hom(N, \mathbf{Q}/\mathbf{Z})$. The cup-product

$$H^1(k_v, S) \times H^1(k_v, M) \longrightarrow \mathrm{Br}(k_v) \xrightarrow{\mathrm{inv}_v} \mathbf{Q}/\mathbf{Z}$$

defines an isomorphism (local Tate duality) of finite groups $H^1(k_v, S) = H^1(k_v, M)^*$ ([Milne, ADT], I.2.3).

Consider the exact sequence (cf. [Milne, ADT], I.4.20 (b), where the statement should be corrected using [Milne, ADT], I.4.13)

$$0 \to \mathrm{III}^1(k, S) \to H^1(k, S) \to P^1(k, S) \to H^1(k, M)^* \tag{6.7}$$

where $P^1(k, S)$ is the restricted product of $H^1(k_v, S)$ for all places v with respect to subgroups $H^1(\mathcal{O}_v, S_v)$ defined for $v \notin T$, that is, for almost all v. The right arrow in (6.7) is the sum of

$$\tau_v : H^1(k_v, S) = H^1(k_v, M)^* \to H^1(k, M)^*$$

which is the composition of the local duality isomorphism and the dual of the restriction map. Any element of $H^1(k, M)$ for almost all places v restricts to an element of $H^1(\mathcal{O}_v, M) \subset H^1(k_v, M)$ ([Milne, ADT], I.4.8), a group orthogonal to $H^1(\mathcal{O}_v, S_v)$ with respect to the local duality pairing. Thus the last arrow in (6.7) is given by a finite sum and hence is well defined.

Let Y be an X-torsor under S of type λ whose existence was established in (1). Let $Y(P)$ denote the element in $H^1(k(P), Y)$ given by the fibre of Y at a closed point P. The map $X(\mathbb{A}_k) \to \prod_{v \in \Omega} H^1(k_v, S)$ sending $\{P_v\}$ to $\{Y(P_v)\}$ has its image in $P^1(k, S)$. Indeed, we can enlarge the set of places T depending on X, Y, and $\{P_v\}$, such that there exists an integral, regular model $\mathcal{X}/Spec(\mathcal{O}_{k,T})$ and an $\mathcal{X}$-torsor $\mathcal{Y}$ under $\mathcal{S}$ which give X, S, and Y at the generic point $Spec(k)$ of $Spec(\mathcal{O}_{k,T})$. Moreover, we can assume that $P_v \in \mathcal{X}(\mathcal{O}_v)$ for $v \notin T$, thus $\mathcal{Y}(P_v) \in H^1(\mathcal{O}_v, S_v)$ for such v.

Thus we can define a map

$$X(\mathbb{A}_k) \to H^1(k, M)^*, \text{ by } \{P_v\} \mapsto \sum_{v \in \Omega} \tau_v(Y(P_v)).$$

From Corollary 4.1.2 it follows that for any $A \in \mathrm{Br}_\lambda(X)$ and any $\alpha \in H^1(k, M)$ such that $r(A) = \lambda_*(\alpha)$ there exists $A_0 \in \mathrm{Br}(k)$ such that for any $P_v \in X(k_v)$ we have

$$A(P_v) = res_{k,k_v}(\alpha) \cup Y(P_v) + res_{k,k_v}(A_0) \in \mathrm{Br}(k_v).$$

This equality combined with global reciprocity implies that

$$\sum_{v \in \Omega} \mathrm{inv}_v(A(P_v)) = \sum_{v \in \Omega} \tau_v(Y(P_v))(\alpha), \qquad (6.8)$$

which is just the Brauer–Manin pairing. Let us now complete the proof of (2). If $\{P_v\} \in X(\mathbb{A}_k)$ is Brauer–Manin orthogonal to $\mathrm{Br}_\lambda(X)$, then we see that $\{Y(P_v)\}$ goes to 0 under the right arrow of (6.7), thus is the image of some $\sigma \in H^1(k, S)$. Note that the image of σ in $P^1(k, S)$ thus belongs to the subgroup $P_T^1(k, S) := \prod_{v \in T} H^1(k_v, S) \times \prod_{v \notin T} H^1(\mathcal{O}_v, \mathcal{S}_v)$. Hence $\{Y^\sigma(P_v)\} = 0 \in P_T^1(k, S)$, in other words, $\{Y^\sigma(P_v)\}$ is the trivial torsor under S over the ring $\prod_{v \in T} k_v \times \prod_{v \notin T} \mathcal{O}_v$. Thus the fibre of $f^\sigma : Y^\sigma \to X$ over $\{P_v\}$ contains a point over this ring, this is an adelic point on Y^σ. This proves (2).

Proof of (3). Let $\{P_v\} \in X(\mathbb{A}_k)$ be the image of an adelic point on Y. Then $Y(P_v) = 0$, and (6.8) implies that $\{P_v\}$ is contained in $X(\mathbb{A}_k)^{\mathrm{Br}_\lambda}$. This is statement (3); the proof of Theorem 6.1.2 *(a)* is now finished. QED

6.2 Manin obstruction and global duality pairings

In this section we prove that three most important classes of algebraic groups, namely, tori, semisimple groups, and abelian varieties, share the property that for torsors X under such groups the Manin obstruction related to $\mathrm{B}(X)$ is the only obstruction to the Hasse principle (for abelian varieties this result is conditional on the finiteness of Ш).

Torsors under tori.

If a k-variety X is such that $\overline{X}$ has non-constant invertible functions, then we cannot directly apply Theorem 6.1.2 to X, but should consider its partial compactification X_c satisfying the property $\overline{k}[X_c]^* = \overline{k}^*$. However, for k-torsors under k-tori a direct approach is available.

Recall that for any k-torsor under a torus T there is a canonical isomorphism of Γ_k-modules $\overline{k}[X]^*/\overline{k}^* = \hat{T}$ (Rosenlicht's lemma), and the class of the extension

$$1 \to \overline{k}^* \to \overline{k}[X]^* \xrightarrow{\gamma} \overline{k}[X]^*/\overline{k}^* = \hat{T} \to 0 \qquad (6.9)$$

in $Ext_k^1(\hat{T}, \overline{k}^*) = H^1(k, T)$ coincides with the class $-[X]$ (Lemma 2.4.3).

The algebraic part of the Brauer group of X is easy to compute using the usual Hochschild–Serre spectral sequence $H^p(k, H^q(\overline{X}, \mathbf{G}_m)) \Rightarrow$

$H^{p+q}(X, \mathbf{G}_m)$. Since $Pic(\overline{X}) = Pic(\overline{T}) = 0$, it immediately gives an isomorphism

$$\mathrm{Br}_1(X) = H^2(k, \overline{k}[X]^*).$$

From (6.9) we get an exact sequence

$$\mathrm{Br}(k) \to \mathrm{Br}_1(X) \to H^2(k, \hat{T}) \to H^3(k, \overline{k}^*).$$

Let k be a number field. Then $H^3(k, \overline{k}^*) = 0$, and we get an isomorphism

$$i : \mathrm{B}(X)/\mathrm{Br}_0(X) \xrightarrow{\sim} \mathrm{III}^2(k, \hat{T}).$$

We have the following result whose proof is very similar to (but simpler than) that of Theorem 6.1.2, the only new ingredient being the result on tori above quoted. Let

$$\langle \cdot, \cdot \rangle : \ \mathrm{III}^1(k, T) \times \mathrm{III}^2(k, \hat{T}) \to \mathbf{Q}/\mathbf{Z}$$

be the global Poitou–Tate pairing. Its definition was recalled in Step 3 of the proof of (1) in Theorem 6.1.2 (a).

Theorem 6.2.1 *Let k be a number field, and X be a k-torsor under a torus T, such that $X(\mathbb{A}_k) \neq \emptyset$. The class of X in $H^1(k, T)$ belongs to the Tate–Shafarevich group $\mathrm{III}^1(k, T)$. Let $A \in \mathrm{B}(X)$ and $\{P_v\} \in X(\mathbb{A}_k)$. Then we have*

$$\sum_{v \in \Omega} \mathrm{inv}_v(A(P_v)) = -\langle [X], i(A) \rangle.$$

The Manin obstruction to the Hasse principle attached to $\mathrm{B}(X)$ is the only obstruction for torsors under tori.

Proof. We shall adapt the proof of part (1) of Theorem 6.1.2 (a) and we keep the same notation.

In view of the canonical isomorphism of Rosenlicht's lemma we shall write $\hat{T} = \overline{k}[X]^*/\overline{k}^* = \overline{k_v}[X]^*/\overline{k_v}^*$.

Choose $\overline{P} \in X(\overline{k})$. Let $e_{\overline{P}}$ be the homomorphic section of the natural injection $\overline{k}^* \hookrightarrow \overline{k}[X]^*$ given by $g \mapsto g(\overline{P})$. When $\overline{P} = P \in X(k)$ the section e_P is Galois equivariant.

Let γ be the map $\overline{k}[X]^* \to \hat{T}$ in (6.9). Let $\sigma_{\overline{P}} : \hat{T} \to \overline{k}[X]^*$ be given by $\sigma_{\overline{P}}(x) = g/g(\overline{P})$, where $\gamma(g) = x$ (such a g is unique up to multiplication by a constant). Then $\sigma_{\overline{P}}$ is a homomorphic section of γ. Note that if $\overline{P}'$ is another $\overline{k}$-point of X, then $\sigma_{\overline{P}}(\gamma(g))/\sigma_{\overline{P}'}(\gamma(g)) \in \overline{k}^*$ equals $g(\overline{P}')/g(\overline{P})$.

The coboundary of $\sigma_{\overline{P}} \in C^0(\Gamma_k, Hom(\hat{T}, \overline{k}[X]^*))$ can be viewed as a 1-cocycle $d\sigma_{\overline{P}} \in Z^1(\Gamma_k, Hom(\hat{T}, \overline{k}^*))$. The class of $d\sigma_{\overline{P}}$ in $Ext_k^1(\hat{T}, \overline{k}^*)$

is by definition the class of extension (6.9). By Lemma 2.4.3 quoted at the beginning of this section, it equals $-[X]$. Note that (in the additive notation) we have $\sigma_{\overline{P}}(\gamma(\cdot)) + e_{\overline{P}}(\cdot) = 1$, hence $d\sigma_{\overline{P}}$ is the inverse of $de_{\overline{P}}$.

Let us consider the global Poitou–Tate pairing $\langle [X], i(A) \rangle$. Let A be given by a 2-cocycle $f = (f_{s,t})$ in $Z^2(\Gamma_k, \overline{k}[X]^*)$. Define the cochain $h \in C^2(\Gamma_k, \overline{k}^*)$ (the global ingredient of the pairing) by $h = -e_{\overline{P}} \cup f = f(\overline{P})^{-1}$. Then $dh = -de_{\overline{P}} \cup f = d\sigma_{\overline{P}} \cup f$.

Let us construct the local factor of the pairing at the place v. Let $\overline{k}_v$ be the algebraic closure of k_v, $\Gamma_v \subset \Gamma_k$ be the decomposition group isomorphic to $Gal(\overline{k}_v/k_v)$. Since P_v is a k_v-point, the homomorphism σ_{P_v} is Γ_v-equivariant. Define a homomorphism $\theta_v : \hat{T} \to \overline{k}_v^*$ by $\theta_v(\gamma(g)) = g(P_v)/g(\overline{P})$. Since σ_{P_v} is Γ_v-equivariant, we have $d\sigma_{P_v} = 0$, and therefore $d\theta_v \in Z^1(\Gamma_v, Hom(\hat{T}, \overline{k}_v^*))$ is the restriction to k_v of the cocycle $d\sigma_{\overline{P}}$ whose class in $H^1(k, Hom(\hat{T}, \overline{k}^*))$ is $-[X]$.

Following the definition of the global Poitou–Tate pairing we define $\epsilon_v \in \mathrm{Br}(k_v)$ as the class of the cocycle $\theta_v \cup \gamma(f) - h = \theta_v \cup \gamma(f) + e_{\overline{P}} \cup f$. The first summand (in the additive notation) equals $f(P_v)/f(\overline{P})$, and the second one $f(\overline{P})$. Therefore, $\epsilon_v \in \mathrm{Br}(k_v)$ is the class of the 2-cocycle $f(P_v) = (f_{s,t}(P_v))$. This proves that $\mathrm{inv}_v(\epsilon_v) = \mathrm{inv}_v(A(P_v))$. This finishes the proof of the displayed formula in the theorem. The last statement follows from the non-degeneracy of the Poitou–Tate duality pairing and the fact that i is an isomorphism. QED

Torsors under semisimple groups.

Let G be a semisimple k-group. Let G^{sc} be the universal covering of G; this is a semisimple simply connected k-group which is a central extension

$$1 \to \mu \to G^{sc} \to G \to 1$$

with some finite Γ_k-module μ. We noted at the end of Section 3.2 that the map $G^{sc} \to G$ is a universal torsor. Let $\delta : H^1(k, G) \to H^2(k, \mu)$ be the differential in the exact sequence of cohomology sets. Let X be a k-torsor under G, and $[X] \in H^1(k, G)$ be its class. Then $\delta([X])$ coincides with the elementary obstruction $e(X)$ (Proposition 3.2.2). This is the obstruction for the existence of universal torsors on X, and in our previous notation it is just $b_X \in H^2(k, \mu)$.

We shall use the known fact that $Pic(\overline{X})$ is canonically isomorphic to $Pic(\overline{G}) = \hat{\mu}$ as a Γ_k-module ([Sansuc 81], 6.7).

When k is a number field we have $H^3(k, \overline{k}^*) = 0$, and hence there is an exact sequence

$$\mathrm{Br}_0(X) \to \mathrm{Br}_1(X) \to H^1(k, Pic(\overline{X})) \to 0.$$

This implies that $\text{III}^1(k, Pic(\overline{X})) = Б(X)/\text{Br}_0(X)$. In our case we obtain an isomorphism

$$i : Б(X)/\text{Br}_0(X) \xrightarrow{\sim} \text{III}^1(k, \hat{\mu}).$$

(For the sake of completeness recall that $\text{Br}(\overline{G}) = 0$ [Iversen].)

In the theorem below $\langle \cdot, \cdot \rangle$ stands for the global Poitou–Tate pairing

$$\text{III}^2(k, \mu) \times \text{III}^1(k, \hat{\mu}) \to \mathbf{Q}/\mathbf{Z}.$$

Theorem 6.2.2 (Sansuc) *Let k be a number field, and X be a k-torsor under a semisimple k-group G, such that $X(\mathbb{A}_k) \neq \emptyset$. Let $A \in Б(X)$ and $\{P_v\} \in X(\mathbb{A}_k)$. Then we have*

$$\sum_{v \in \Omega} \text{inv}_v(A(P_v)) = \langle \delta([X]), i(A) \rangle.$$

The Manin obstruction attached to $Б(X)$ is the only obstruction to the Hasse principle for torsors under semisimple groups.

Proof. The displayed formula is simply (6.4) in view of the identification $b_X = \delta([X])$ explained above. Let us deduce the last claim from the displayed formula. The non-degeneracy of the Poitou–Tate pairing and the fact that i is an isomorphism imply that if there is no obstruction attached to $Б(X)$ then $\delta([X]) = 0$. Let Y be a k-torsor under G^{sc} such that the class $[X] \in H^1(k, G)$ is the image of the class $[Y] \in H^1(k, G^{sc})$ under the natural map $H^1(k, G^{sc}) \to H^1(k, G)$. (Note, by the way, that Y is a universal X-torsor.) Since μ is central in G^{sc} the twisting of a k-torsor under G^{sc} by a k-torsor under μ defines an action of the group $H^1(k, \mu)$ on the set $H^1(k, G^{sc})$. (In general, we could get a k-torsor under an inner form of G^{sc}.) The class $[Y]$ is unique up to this action of $H^1(k, \mu)$. Since $X(k_v) \neq \emptyset$ for all $v \in \Omega$, the image of $[Y] \in H^1(k, G^{sc})$ in $H^1(k_v, G^{sc})$ comes from $H^1(k_v, \mu)$. Let us recall a well known fact, that the natural map

$$H^1(k, \mu) \to \prod_{v \in \Omega_\infty} H^1(k_v, \mu),$$

where Ω_∞ is the set of infinite places of k, is surjective (see, e.g., [PR], Section VII.3, Cor. 2). Modifying $[Y]$ by the image of an element of $H^1(k, \mu)$ which has the same restrictions at the infinite places as $[Y]$, we ensure that Y is such that $Y(k_v) \neq \emptyset$ for all $v \in \Omega_\infty$. It is a fundamental fact in the arithmetic theory of semisimple simply connected groups that the Hasse principle holds in the following strong form:

$$H^1(k, G^{sc}) = \prod_{v \in \Omega_\infty} H^1(k_v, G^{sc}).$$

(Kneser, Harder and Chernousov; see [PR], Section VI.1, Thms. 4 and 6. In fact, $H^1(k_v, G^{sc}) = 1$ for any finite place v.) This implies that $Y(k) \neq \emptyset$, hence $X(k) \neq \emptyset$. QED

More generally, the last statement of Theorems 6.2.1 and 6.2.2 remains true for all connected linear groups [Sansuc 81].

Torsors under abelian varieties.

Finally, we note that a similar formula is available for abelian varieties. If S is an abelian variety we denote by S^t its dual abelian variety, $S^t(\overline{k}) = Pic^0(\overline{S})$. If X is a k-torsor under S, then we have a canonical isomorphism of Γ_k-modules $Pic^0(\overline{X}) = Pic^0(\overline{S})$. This gives a map

$$j : \text{III}(S^t) \to \text{III}^1(k, Pic(\overline{X})) = \text{Б}(X)/\text{Br}_0(X).$$

In the following theorem we denote by $\langle \cdot, \cdot \rangle$ the Cassels–Tate pairing

$$\text{III}(S) \times \text{III}(S^t) \to \mathbf{Q}/\mathbf{Z}.$$

Let us recall the definition of $\langle \xi, \alpha \rangle$ ([Milne, ADT], Remark I.6.11). We realize $\xi \in \text{III}(S)$ as the class of a k-torsor X under S. Consider the exact sequence of Γ_k-modules

$$1 \to \overline{k}(X)^*/\overline{k}^* \to Div(\overline{X}) \to Pic(\overline{X}) \to 0.$$

Let $\alpha \in \text{III}(S^t)$. Then $\delta(j(\alpha)) \in \text{III}^2(k, \overline{k}(X)^*/\overline{k}^*)$. Since $H^3(k, \overline{k}^*) = 0$, this element can be lifted to $f \in H^2(k, \overline{k}(X)^*)$ with the property that the restriction of f to k_v comes from $f_v \in H^2(k_v, \overline{k_v}^*)$. Then $\langle \xi, \alpha \rangle$ is defined as $\sum_{v \in \Omega} \text{inv}_v(f_v)$. Note that one can compute $\text{inv}_v(f_v)$ by evaluating f in a k_v-point of X that avoids $div(f)$.

Theorem 6.2.3 (Manin) *Let k be a number field, and X be a k-torsor under an abelian variety S, such that $X(\mathbb{A}_k) \neq \emptyset$. The class of X in $H^1(k, S)$ belongs to the Tate–Shafarevich group $\text{III}(S)$. Let $A \in \text{Б}(X)$ be such that its image $r(A) \in \text{Б}(X)/\text{Br}_0(X)$ equals $j(a)$ for some $a \in \text{III}(S^t)$. Then for any $\{P_v\} \in X(\mathbb{A}_k)$ we have*

$$\sum_{v \in \Omega} \text{inv}_v(A(P_v)) = -\langle [X], a \rangle.$$

For an abelian variety S with the property that $\text{III}(S)$ is finite, the Manin obstruction attached to $\text{Б}(X)$ is the only obstruction to the Hasse principle for k-torsors under S.

The last statement of Theorem 6.2.3 is more precise than the one given in the example after Corollary 6.1.3 ($\text{III}(S)$ is conjecturally finite, whereas $H^1(k, S)$ is certainly not).

Proof of theorem. The deduction of Theorem 6.2.3 from the definition of the Cassels–Tate pairing given above is immediate from Step 0 of the proof of (1) in Theorem 6.1.2 *(a)*. QED

Let us describe the closed set $S(\mathbb{A}_k)^{\text{Br}} \subset S(\mathbb{A}_k)$. It must contain the closure of the diagonal image of $S(k)$. On the other hand, when $k_v = \mathbf{R}$ the continuous function $\text{inv}_v(A(P_v))$ with values in $\{0, \frac{1}{2}\}$ must be constant on each connected component of $A(\mathbf{R})$. The product of the connected components of 0, when k_v is archimedean, and the one-element set $\{0\}$, when k_v is non-archimedean, is the connected component of 0 in $S(\mathbb{A}_k)$. We conclude that $S(\mathbb{A}_k)^{\text{Br}}$ contains the subgroup generated by the closure of $S(k)$ and the connected component of $S(\mathbb{A}_k)$. In fact, this is all.

Proposition 6.2.4 *Suppose that $\text{III}(S)$ is finite; then $S(\mathbb{A}_k)^{\text{Br}}$ coincides with the subgroup of $S(\mathbb{A}_k)$ generated by the closure of the diagonal image of $S(k)$ and the connected component of $S(\mathbb{A}_k)$.*

Proof. Since S and S' are isogenous, the finiteness of $\text{III}(S)$ implies the finiteness of $\text{III}(S')$. Consider the diagonal embedding

$$S(k) \subset \prod_{v \in \Omega} H^0(k_v, S).$$

Here we understand that for an archimedean place v, $H^0(k_v, S)$ is the group of connected components of $S(k_v)$, and $H^0(k_v, S) = S(k_v)$ otherwise. Under the finiteness assumption for the Tate–Shafarevich group, the global duality theory ([Milne, ADT], I.6) describes the kernel of the pairing of $\prod_{v \in \Omega} H^0(k_v, S)$ with $H^1(k, S^t)$ defined as the sum of local Tate duality pairings

$$H^0(k_v, S) \times H^1(k_v, S^t) \to \mathbf{Q}/\mathbf{Z}.$$

It is a profound result that this kernel is precisely the closure of the diagonal image of $S(k)$ (this is the beginning of the so called 'dual sequence of Cassels'; see [Milne, ADT], p. 102). By a theorem of Serre it is isomorphic to the profinite completion of $S(k)$.

We identify $\text{Br}_1(S)$ with the direct sum $\text{Br}(k) \oplus H^1(k, Pic(\overline{S}))$, so that to $\xi \in H^1(k, Pic(\overline{S}))$ there corresponds an element $A_\xi \in \text{Br}_1(S)$ such that $r(A_\xi) = \xi$, normalized by the condition $A_\xi(0) = 0$. Considering $S^t(\overline{k})$ as $Pic^0(\overline{S})$, we obtain another pairing between $H^0(k_v, S)$ and $H^1(k_v, S^t)$, namely, $\mathrm{inv}_v(A_\xi(P_v)) \in \mathbf{Q}/\mathbf{Z}$, where ξ comes from $H^1(k, S^t)$, and P_v represents an element of $H^0(k_v, S)$. It is known that these two pairings coincide ([Manin 70], Prop. 8 (c)). Thus the kernel of the Brauer–Manin pairing is contained in the kernel of the global duality paring. By the remarks before the proposition we also have the opposite inclusion. QED

This describes the situation with weak approximation on abelian varieties: the Brauer–Manin obstruction to weak approximation away from Ω_∞ is the only one. In other words, the failure of weak approximation is 'almost' controlled by the Brauer–Manin obstruction, the only problem being the archimedean places. See [Wang] for more on this subject.

Curves.

The comparison of the Brauer–Manin pairing with the global duality pairings on abelian varieties has some immediate corollaries for the theory of the Manin obstruction on curves.

Corollary 6.2.5 *Let C be a smooth and proper curve over a number field k such that $C(\mathbb{A}_k) \neq \emptyset$, but C has no divisor class of degree 1 defined over k, in particular, $C(k) = \emptyset$. Suppose that $\mathrm{III}(J)$ is finite, where J is the Jacobian of C. Then C is a counter-example to the Hasse principle explained by the Manin obstruction.*

Proof. Let X be the k-torsor under J such that for any field extension L/k the L-points of X parametrize divisor classes of degree 1 on C defined over L. It can be obtained by twisting J by the 1-cocycle of Γ_k with coefficients in $J(\overline{k})$ given by $\gamma \mapsto {}^\gamma \overline{x}_0 - \overline{x}_0$ where $\overline{x}_0$ is some $\overline{k}$-point on C. Then the $\overline{k}$-morphism $\overline{C} \to \overline{J}$ sending x to $x - \overline{x}_0$ descends to a k-morphism $\phi : C \hookrightarrow X$ (cf. [Weil 55]). It follows from our assumption that $X(k) = \emptyset$, but X has points everywhere locally since the same is true for C. Hence X is a counter-example to the Hasse principle. By Theorem 6.2.3 we have $X(\mathbb{A}_k)^{\mathrm{Br}} = \emptyset$. According to (5.3) the morphism ϕ maps $C(\mathbb{A}_k)^{\mathrm{Br}}$ to $X(\mathbb{A}_k)^{\mathrm{Br}}$. Hence $C(\mathbb{A}_k)^{\mathrm{Br}}$ is empty. QED

There are many examples of this kind. Such is the quartic curve $x^4 + y^4 = 241z^4$ over $k = \mathbf{Q}$ (Bremner and Cassels, see [Cassels 85a], Sect. 3), it has no rational points in any algebraic number field of odd degree. Here the group of $\mathbf{Q}$-points of the Jacobian is finite. There are also more complicated examples when this group is infinite (*op. cit.*, Sect. 4).

Corollary 6.2.6 (Scharaschkin [Sch]) *Let C be a smooth and proper*

curve over a number field k such that $C(\mathbb{A}_k) \neq \emptyset$, but $C(k) = \emptyset$. Assume that $\text{III}(J)$ is finite, where J is the Jacobian of C. If $J(k)$ is finite, then C is a counter-example to the Hasse principle explained by the Manin obstruction.

Proof. As in the proof of Corollary 6.2.5 we realize C as a subset of a principal homogeneous space X under J. If $X(k) = \emptyset$ we conclude the proof by Corollary 6.2.5. In the opposite case X is isomorphic to J, so that $C \subset J$. By Proposition 6.2.4 we have that $C(\mathbb{A}_k)^{\text{Br}}$ is contained in the subgroup of $J(\mathbb{A}_k)$ generated by the connected component of 0, and the diagonal image of $J(k)$ (this set is finite by assumption, and hence coincides with its closure). Let $\{P_v\} \in C(\mathbb{A}_k)^{\text{Br}}$. Then for non-archimedean v we have $P_v \in C(k_v) \cap J(k) = C(k)$. Since $C(k) = \emptyset$ by assumption, we must also have $C(\mathbb{A}_k)^{\text{Br}} = \emptyset$. QED

In some counter-examples to the Hasse principle on curves covered neither by Corollary 6.2.6 nor by Corollary 6.2.5 the absence of rational points can still be explained by the Manin obstruction. Let us consider the plane curve C over $k = \mathbf{Q}$ given by the equation

$$C : \quad 3x^4 + 4y^4 = 19z^4.$$

This is a counter-example to the Hasse principle. As an additional feature C has a point over a cubic extension of $\mathbf{Q}$, hence a 0-cycle of degree 1 [BLM]. It can be used to realize C as a subvariety of its Jacobian J. The Jacobian J is isogenous to the product of three elliptic curves of positive rank, thus k-points are Zariski dense in J.

An argument of Cassels can be used to show that the Manin obstruction suffices to explain this counter-example. Cassels observes ([Cassels 85a], Sect. 5, [Cassels 85b]) that the equation of C can be rewritten as

$$(19z^2 - 8y^2 - 3x^2)(19z^2 + 8y^2 - 3x^2) = 57(x^2 - z^2)^2.$$

Consider the rational function $f = (19z^2 - 8y^2 - 3x^2)/x^2$. One easily checks using this factorization that $div(f)$ is divisible by 2 in $Div(\overline{C})$. Hence the normalization of C in $\mathbf{Q}(C)(\sqrt{f})$ is an unramified double covering $Y \to C$. Then Cassels shows that for any twist $Y' \to C$ of $Y \to C$ by a k-torsor under $\mathbf{Z}/2$ (in down-to-earth terms this is simply the covering of C given by $\sqrt{af}$, for some $a \in \mathbf{Q}^*$) we have $Y'(\mathbb{A}_\mathbf{Q}) = \emptyset$. By Theorem 6.1.2 we conclude that $C(\mathbf{Q})^{\text{Br}} = \emptyset$. For a different approach see [Sch].

I do not know whether or not there exists a smooth and projective curve over a number field, which is a counter-example to the Hasse principle not explained by the Manin obstruction.

Diagonal cubic surfaces.

Let us mention another recent result which depends on the finiteness conjecture for the Tate–Shafarevich group.

Theorem 6.2.7 (Heath-Brown [HB], [BF]) *Let p_1, p_2, p_3, p_4 be primes congruent to 2 modulo 3. Suppose that the Tate–Shafarevich groups of elliptic curves over $\mathbf{Q}$ are finite. Then the cubic surface given in $\mathbf{P}^3_{\mathbf{Q}}$ by the equation*

$$p_1 X_1^3 + p_2 X_2^3 + p_3 X_3^3 + p_4 X_4^3 = 0$$

has a $\mathbf{Q}$-rational point.

Note that such a surface has points everywhere locally; moreover, it is known that the Manin obstruction on it is empty. More precisely, if a prime number divides exactly one coefficient of the equation of a diagonal cubic surface, then there is no Manin obstruction to the Hasse principle ([CKS], Corollaire, p. 3; the coefficients are assumed to be cube free). An example of the opposite situation is the generalized counter-example of Bremner ([CKS], Prop. 5): cubic surfaces given by

$$X_1^3 + p^2 X_2^3 + pq X_3^3 + q^2 X_4^3 = 0$$

where p and q are prime numbers congruent modulo 9 to 2 and 5, respectively, are counter-examples to the Hasse principle. See [CKS], Sect. VII, for the analysis of this and other examples. A short description of the algorithm computing the Manin obstruction to the Hasse principle on diagonal cubic surfaces can be found in [C86].

The proof of Theorem 6.2.7 in [HB] relies on an unproved conjecture of Selmer. T. Fisher showed how to circumvent this difficulty. The detailed exposition of this corrected proof can be found in [BF]. Very recently, Swinnerton-Dyer combined this with his powerful method ([SD95], further developed in [CSS98b], [BSw] and [SD00a]) and proved a much stronger theorem.

Theorem 6.2.8 (Swinnerton-Dyer [SD00b]) *Let k be a number field not containing the primitive cube roots of unity. Assume that the Tate–Shafarevich group of every elliptic curve $X^3 + Y^3 = AZ^3$ over any quadratic extension of k is finite. If the equation*

$$a_1 X_1^3 + a_2 X_2^3 + a_3 X_3^3 + a_4 X_4^3 = 0$$

is everywhere locally soluble, then each of the following three criteria is sufficient for its solubility in k.

(i) There exist primes v_1, v_3 of k not dividing 3 such that a_1 is a non-unit at v_1 and a_3 is a non-unit at v_3, but for $j = 1$ or 3 the three a_i with $i \neq j$ are units at v_j.

(ii) There is a prime v of k not dividing 3 such that a_1 is a non-unit at v but the other a_i are units there; and a_2, a_3, a_4 are not all in the same coset of k_v^{*3}.

(iii) There is a prime v of k not dividing 3 such that exactly two of the a_i are units at v, and the cubic surface is not birationally equivalent to a plane over k_v.

The condition in this theorem is local; it is slightly stronger than the disappearance of the Manin obstruction. From this remarkable theorem Swinnerton-Dyer deduces the following corollary.

Corollary 6.2.9 *Assume that k does not contain the primitive cube roots of unity and that the Tate–Shafarevich group of every elliptic curve $X^3 + Y^3 = AZ^3$ over any quadratic extension of k is finite. If $b_1, \ldots, b_5$ are non-zero elements of k such that*

$$b_1 X_1^3 + b_2 X_2^3 + b_3 X_3^3 + b_4 X_4^3 + b_5 X_5^3 = 0$$

is everywhere locally soluble, then it is soluble in k.

6.3 Compactifications of torsors under tori

Historically, one of the first successes of the descent method with a torus as the structure group was the class of torus compactifications. More precisely, we have the following result.

Theorem 6.3.1 (Colliot-Thélène–Sansuc) *Consider the class of smooth proper varieties X over a number field k such that X contains a k-torsor under a torus as a dense open set. The Manin obstruction to the Hasse principle and weak approximation is the only obstruction for this class.*

Proof. Let $j : U \hookrightarrow X$ be a dense open set which is a k-torsor under a torus R. We have $Pic(\overline{U}) = 0$, $\overline{k}[U]^*/\overline{k}^* = \hat{R}$ (Lemma 2.4.3), hence there is an exact sequence of Γ_k-modules

$$0 \to \hat{R} \to Div_{\overline{X} \setminus \overline{U}}(\overline{X}) \to Pic(\overline{X}) \to 0.$$

Since $Div_{\overline{X} \setminus \overline{U}}(\overline{X})$ is a finitely generated abelian group, then so is $Pic(\overline{X})$. Let S be the torus dual to the Γ_k-module $Pic(\overline{X})$. The exact sequence of tori dual to the previous exact sequence of Γ_k-modules

$$1 \to S \to M \to R \to 1$$

satisfies the properties (1) and (2) of Section 4.3. As in that section we denote by $\mathcal{M}_R$ the R-torsor under S given by this exact sequence.

Suppose that $X(\mathbb{A}_k)^{\mathrm{B}(X)}$ is not empty. By Proposition 6.1.4 universal X-torsors exist; let T be a universal X-torsor. By Theorem 4.3.1, $j^*(T)$ is a U-torsor under S obtained from $\mathcal{M}_R$ by the base change $\phi : U \to R$ such that the map $\hat{\phi} : \hat{R} = \overline{k}[U]^*/\overline{k}^* \to \overline{k}[U]^*$ is a lifting of $-Id$. By Lemma 2.4.3 the existence of such a lifting implies that U is the trivial k-torsor R, hence $U(k) \neq \emptyset$. We have found a k-point on X.

Fixing a k-point on U we identify U with R. Then $-\phi$ is a translation by a k-point x of R. It follows that $j^*(T)$ is the twisted torsor $(\mathcal{M}_R)^\sigma$ for some $\sigma \in H^1(k, S)$. (This σ is the image of x under the map $R(k) \to H^1(k, S)$; it is the same thing as the class of the fibre of $\mathcal{M}_R$ at x.) But $(\mathcal{M}_R)^\sigma$ is none other than the k-torsor under M whose class is the image of σ in $H^1(k, M)$. Recall that M is a quasi-trivial torus. From Hilbert's Theorem 90 it follows that $H^1(k, M) = 0$. Hence $(\mathcal{M}_R)^\sigma$ is isomorphic to M as a k-variety. On the other hand, M is a dense open subset of an affine space. Thus T is k-rational, hence satisfies weak approximation. It remains to apply Corollary 6.1.3. QED

Note that the proof shows that the Manin obstruction to the Hasse principle associated to $\mathrm{B}(X)$ is the only obstruction (cf. Proposition 6.1.4). For weak approximation we need the whole of $\mathrm{Br}(X)$ (to be able to lift an adelic point on X to an adelic point on some universal torsor).

The above argument shows that universal torsors with a k-point are k-rational. This works over any field k. If k is a number field, only finitely many universal torsors of a given type contain k-points. We conclude that the set $X(k)$ is divided into a finite disjoint union of subsets parametrized by rational functions. For more details, and connections with R-equivalence, see [CS77].

Theorem 6.3.2 (T. Ono, V.E. Voskresenskiĭ) *Smooth proper surfaces over a number field k containing a k-torsor under a torus as a dense open set satisfy the Hasse principle and weak approximation.*

Proof. We shall compute that $H^1(k, Pic(\overline{X})) = 0$; then this theorem becomes a corollary of the previous one. To do this we may assume that X contains a k-point. Indeed, let $K = k(X)$. Then $X_K = X \times_k Spec(K)$ has a K-point given by the 'diagonal' image of the generic point $Spec(K) \hookrightarrow X$.

On the other hand, since $Pic(\overline{X})$ is a free abelian group of finite type, the group $H^1(k, Pic(\overline{X}))$ depends only on the image of Γ_k in the automorphism group of $Pic(\overline{X})$. This follows from the 'restriction–inflation' sequence and the fact that H^1 of a profinite group with coefficients in the trivial module **Z** is 0.

It is well known that $H^1(k, Pic(\overline{X}))$ is a birational invariant of X ([Manin, CF], IV, Thm. 1.3). (The geometric Picard group of a blowing up of X is the direct sum of $Pic(\overline{X})$ and a permutation Γ_k-module generated by the classes of the exceptional curves. Every birational transformation can be decomposed into a sequence of blowings up of smooth points, and the inverses of such maps. Finally, the first cohomology of a permutation module is trivial.)

V.E. Voskresenskiĭ has proved (by inspection) that all tori of dimension 1 and 2 are k-rational ([V, AT], IV.9, Thms. 4.73, 4.74; see also [Manin, CF], IV.8). Hence X is also k-rational as long as it contains a rational point. This completes the proof. QED

Another proof of this result can be found in [V, AT], VI.7.

Corollary 6.3.3 ([Manin, CF], IV, Thm. 8.6) *Del Pezzo surfaces of degree 6, that is, $(\overline{k}/k)$-forms of $\mathbf{P}^2$ with three non-collinear points blown up, satisfy the Hasse principle and weak approximation.*

Proof. The projective plane with three non-collinear points blown up contains exactly six exceptional curves: three inverse images of blown-up points and the proper transforms of lines passing through two of them. Thus this configuration of exceptional curves is globally defined over k. Let X be a Del Pezzo surface of degree 6, and let $U \subset X$ be the dense open set complementary to the union of six exceptional curves. It is clear that $\overline{U} \simeq \mathbf{G}_m^2$. By Lemma 2.4.4 U is a k-torsor under a 2-dimensional torus. QED

The proof of this fact given here illustrates the different techniques and uses many results presented in this book, but it is certainly not the most direct one.

Comments

Theorem 6.1.1 was proved by Colliot-Thélène and Sansuc in [CS87a] when $Pic(\overline{X})$ is torsion free; the present, more general form is taken from [S99] (see also [HS]). Note that even the long proof of that theorem in [CS87a]

is not completely self-contained (what corresponds there to our Step 0 is proved in the Annexe to [CS77]). Sansuc's general theorem is that the Manin obstruction to the Hasse principle and weak approximation is the only one for principal homogeneous spaces under connected linear groups. This theorem and much more are proved in his profound paper [Sansuc 81]. This paper motivated the subsequent research of Borovoi ([Bor 96], [Bor 98]). We have chosen to prove the most important particular cases of Sansuc's theorem, where simple proofs are available. Manin's theorem about the obstruction of his name on abelian varieties ([Manin 70], [Manin, CF], Ch. VI) was his initial insight the into possible significance of the Brauer group for the Hasse principle. Although it added nothing new to the arithmetic of abelian varieties, the Manin obstruction provided the first conceptual approach to the Hasse principle on arbitrary algebraic varieties.

Let us formulate some open problems related to the material of this chapter.

Question 1. Is the Manin obstruction attached to $B(X)$ the only obstruction to the Hasse principle for k-torsors under arbitrary connected algebraic k-groups?

This is likely to be true. Probably, a much harder question is

Question 2. Is the Manin obstruction the only obstruction to the Hasse principle on smooth and projective curves?

This is true if points are replaced by 0-cycles of degree 1. (For a simple proof of this fact, conditionally on the finiteness of the Tate–Shafarevich group, see [C99], Prop. 3.7.)

7

Abelian descent on conic bundle surfaces

In this chapter we prove that the Manin obstruction is the only obstruction to the Hasse principle and weak approximation for (smooth and projective models of) conic bundle surfaces defined by the equation

$$y^2 - bz^2 = aP(x), \qquad (7.1)$$

provided that $P(x)$ is a monic polynomial of degree 4, or a product of polynomials of degrees 4 and 2. The degree 4 case was proved by Colliot-Thélène, Sansuc and Swinnerton-Dyer in their remarkable paper [CSS87]. The second result was recently obtained by Swinnerton-Dyer [SD99]. The method consists of showing that the torsors defined by the morphism to $\mathbf{P}_k^1$ (the projection to the x-axis) as in Section 4.4, satisfy the Hasse principle and weak approximation, and then concluding the proof by Corollary 6.1.3. In our opinion this constitutes one of the most spectacular applications of descent with torsors under tori on varieties which are far from homogeneous varieties[1]. An easy computation of the Brauer group then shows that the Hasse principle holds provided $P(x)$ is irreducible.

After a rational change of variables we can assume that $n = deg(P(x))$ is even. If $n = 2$ our surface is a quadric, and that the set of its smooth points satisfies the Hasse principle is the original Minkowski–Hasse theorem. F. Châtelet was the first to study the arithmetic of such surfaces when $P(x)$ is a product of four different linear factors [Châtelet]. Whenever $n = 4$ one refers to such surfaces as *Châtelet surfaces*. By Proposition 4.4.8 all we need to do in this case is to show that a certain complete intersection $Y = Y_\alpha$ of two quadrics in $\mathbf{P}_k^7$ satisfies the smooth Hasse principle. The

---

[1] Another approach to this theorem for $deg(P(x)) = 4$ is through Salberger's theorem that the Manin obstruction is the only obstruction for 'the Hasse principle for 0-cycles of degree 1' on conic bundles [Sl88]. Instead of descent this approach uses an ingenious approximation method, but it is not clear whether it can yield Swinnerton-Dyer's result.

equations show that Y contains two skew $\mathbf{P}^3$'s conjugate over $k(\sqrt{b})$. This key observation makes it possible to derive the desired result from the analogous result for smooth intersections of two quadrics in $\mathbf{P}^4_k$ which contain a pair of skew conjugate lines. When $n = 6$ we need to consider a certain complete intersection of four quadrics in $\mathbf{P}^{11}_k$. Its intricate analysis builds upon the analysis of the case $n = 4$.

7.1 Brauer group of conic bundles

Let us explicitly describe a natural smooth and proper model of the surface (7.1). The case when $n = deg(P(x))$ is odd is reduced to the case $deg(P(x))$ even: on replacing x by $x - c$ we can assume that $P(0) \neq 0$; then we set $x = X^{-1}$, $y = X^{-(n+1)/2}Y$, $z = X^{-(n+1)/2}Z$. It is also clear what birational transformation reduces the case when $P(x)$ has repeated factors to the case $P(x)$ separable. Finally, if $P(x)$ has a monic irreducible factor $Q(x)$ such that b is a square in the corresponding residue field, then for some polynomials $A(x)$, $B(x)$ we have[1] $A(x)^2 - bB(x)^2 = Q(x)$. Using the multiplicativity of norms one then easily finds a birational transformation reducing (7.1) to a similar equation whose right hand side is $aP(x)Q(x)^{-1}$.

Now we assume that n is even, $P(x)$ is separable, and that b is not a square in the extensions of k corresponding to the irreducible factors of $P(x)$. Let $P_t(x)$, $t \in T$, be the irreducible factors of $P(x)$ (we use the same notation as in Section 4.4), and let $k_t = k[x]/(P_t(x))$. Let $K_t = k_t(\sqrt{b})$. Over $\overline{k}$ we have $P(x) = (x - e_1)\ldots(x - e_n)$ for some pairwise different $e_i \in \overline{k}$.

Let X_1 (resp. X_2) be the closed subset of $\mathbf{P}^2_k \times (\mathbf{P}^1_k \setminus \infty)$ (resp. of $\mathbf{P}^2_k \times (\mathbf{P}^1_k \setminus 0)$) with coordinates $(y : z : u; x)$ (resp. with coordinates $(Y : Z : U; X)$) given by $y^2 - bz^2 - aP(x)u^2 = 0$ (resp. by $Y^2 - bZ^2 - aX^nP(X^{-1})U^2 = 0$). We glue these two conic bundles over $\mathbf{P}^1_k \setminus (0 \cup \infty)$ by setting $x = X^{-1}$, $y = Y$, $z = Z$, $u = X^{n/2}U$. The resulting surface X is smooth and proper. We shall refer to it as the natural smooth compactification of the affine surface $y^2 - bz^2 = aP(x)$. The singular $\overline{k}$-fibres of the morphism $\pi : X \rightarrow \mathbf{P}^1_k$ are located over the points $x = e_i$. Each such fibre is the union of two smooth rational curves meeting at one point. Note that under our assumptions π has no section: a section meets exactly one component of a singular fibre (otherwise the intersection index

---

[1] Indeed, the prime ideal $(Q(x)) \subset k[x]$ is split in the field extension $k(x)(\sqrt{b})/k(x)$, say $(Q(x)) = I\overline{I}$. Since $k(\sqrt{b})[x]$ is a principal ideal domain, I is generated by a monic polynomial $R(x) \in k(\sqrt{b})[x]$. Write $R(x) = A(x) + B(x)\sqrt{b}$, where $A(x), B(x) \in k[x]$, then $A(x)^2 - bB(x)^2 = Q(x)$.

of the section with the fibre will be greater than 1), but this implies that
b is a square in the residue field of this fibre. In other words, the generic
fibre of π is a conic with no rational point.

Let us now choose a convenient basis in $Pic(\overline{X})$. Fix a square root
$\sqrt{b}$ of b. Let $l \subset \overline{X}$ be the smooth rational curve given by $y = \sqrt{b}$, $z = 1$,
$u = 0$; $l_i \subset \overline{X}$, $i = 1, \dots, n$, be the smooth rational curve given by $y = \sqrt{b}z$,
$x = e_i$; finally, let X_∞ denote the fibre of π at $x = \infty$, that is, $X_\infty = X \backslash X_1$,
this is also a smooth rational curve. We claim that $Pic(\overline{X})$ is generated
by the classes of $X_\infty, l, l_1, \dots, l_n$. Indeed, the following natural complex of
Γ_k-modules is exact:

$$0 \to N \to Pic(\overline{X}) \to Pic(X_{\overline{k}(x)}) \to 0. \qquad (7.2)$$

Here $N \subset Pic(\overline{X})$ is the submodule generated by the components of the
fibres of π, and $X_{\overline{k}(x)}$ is the generic fibre of the morphism $\overline{X} \to \mathbf{P}^1_{\overline{k}}$. The homomorphism $Pic(\overline{X}) \to Pic(X_{\overline{k}(x)})$ given by the restriction to the generic
fibre, is surjective: the Zariski closure of a divisor of $X_{\overline{k}(x)}$ in $\overline{X}$ is a Weil
divisor, but it is also a Cartier divisor since X is smooth. Since l is a
section of $\overline{X} \to \mathbf{P}^1_{\overline{k}}$ and $Pic(X_{\overline{k}(x)}) = \mathbf{Z}$, this last group is generated by the
image of $[l] \in Pic(\overline{X})$. The class of an irreducible component of a fibre of
π is of one of the following types: $[X_\infty]$, $[l_i]$, $[X_\infty] - [l_j]$. This proves our
claim.

In fact we have found more than just a basis: we have found a short
resolution of N, the essential part of $Pic(\overline{X})$, by permutation modules, as
we shall now see. Using it we compute $H^1(k, Pic(\overline{X}))$.

Proposition 7.1.1 *Let k be a perfect field. Let X be the natural smooth
compactification of the surface given by*

$$y^2 - bz^2 = aP(x),$$

*where $P(x) \in k[x]$ is a polynomial of even degree $(a, b \in k^*)$. Let $\overline{d}_1, \dots, \overline{d}_m$
be the residues modulo 2 of the degrees of the irreducible factors of $P(x)$.
Then $H^1(k, Pic(\overline{X}))$ is isomorphic to the quotient of the orthogonal com-
plement to $(\overline{d}_1, \dots, \overline{d}_m)$ in $(\mathbf{Z}/2)^m$ modulo $(1, \dots, 1)$ (with respect to the
standard diagonal quadratic form). In particular, if $P(x)$ is irreducible, or
a product of an irreducible polynomial and a linear factor, then we have
$H^1(k, Pic(\overline{X})) = 0$.*

Proof. Let us first compute $H^1(k, N)$. We employ the same notation
as in Section 4.4. Let $U' \subset \mathbf{P}^1_k$ be the complement to the union of closed

points of $\mathbf{P}^1_k$ corresponding to singular fibres of π and the fibre at infinity. Let $U = \pi^{-1}(U') \subset X$. Since the generic fibre of $\overline{X} \to \mathbf{P}^1_{\overline{k}}$ has no invertible functions other than those coming from the field of functions of $\mathbf{P}^1_{\overline{k}}$, we have $\overline{k}[U]^* = \overline{k}[U']^*$. Then we have an exact sequence of Γ_k-modules:

$$0 \to \overline{k}[U']^*/\overline{k}^* \to Div_{\overline{X}\backslash\overline{U}} \to N \to 0.$$

The left and the middle terms are permutation Γ_k-modules: $Div_{\overline{X}\backslash\overline{U}}$ is freely generated by $X_\infty, l_1, \dots, l_n, l'_1, \dots, l'_n$, where l'_i denotes the component of the fibre at $x = e_i$ other than l_i; and $\overline{k}[U']^*/\overline{k}^*$ is freely generated by the functions $x - e_i$. Their Γ_k-module structure is

$$\overline{k}[U']^*/\overline{k}^* = \bigoplus_{t\in T}\mathbf{Z}[\Gamma_{k_t}/\Gamma_k], \quad Div_{\overline{X}\backslash\overline{U}} = \mathbf{Z} \oplus \bigoplus_{t\in T}\mathbf{Z}[\Gamma_{K_t}/\Gamma_k].$$

The image of $x - e_i$ in $Div_{\overline{X}\backslash\overline{U}}$ is $l_i + l'_i - X_\infty$. By Shapiro's lemma we have $H^1(k, \mathbf{Z}[\Gamma_k/\Gamma_L]) = H^1(L, \mathbf{Z}) = 0$, and $H^2(k, \mathbf{Z}[\Gamma_k/\Gamma_L]) = H^2(L, \mathbf{Z}) = Hom(\Gamma_L, \mathbf{Q}/\mathbf{Z})$. Since K_t is quadratic over k_t, the characters of Γ_{k_t} which are trivial on Γ_{K_t} are of order 2. Identifying $Hom(\Gamma_{k_t}, \mathbf{Z}/2)$ with k_t^*/k_t^{*2} we see that the kernel of the restriction $Hom(\Gamma_{k_t}, \mathbf{Z}/2) \to Hom(\Gamma_{K_t}, \mathbf{Z}/2)$ is generated by the (non-trivial) class of b in k_t^*/k_t^{*2}. A collection of elements in these kernels for all $t \in T$ is in the image of $H^1(k, N)$ if and only if the product of their norms to k is a square in k^* (this condition is given by the 'augmentation' homomorphism $\overline{k}[U']^*/\overline{k}^* \to \mathbf{Z}$, which sends each element of the permutation basis to 1). This is the orthogonality condition of our statement, which computes $H^1(k, N)$.

Note that $H^1(k, Pic(X_{\overline{k}(x)})) = H^1(k, \mathbf{Z}) = 0$, and $Pic(\overline{X})^{\Gamma_k}$ is generated by certain sums of the components of fibres of π, and $l + l'$, where l' is the conjugate divisor of l, given by $y = -\sqrt{b}$, $z = 1$, $u = 0$. Therefore, the cokernel of $Pic(\overline{X})^{\Gamma_k} \to Pic(X_{\overline{k}(x)})^{\Gamma_k}$ is $\mathbf{Z}/2$ (generated by the image of $[l] \in Pic(\overline{X})$). Considering the cohomology of (7.2) we obtain $H^1(k, Pic(\overline{X})) = H^1(k, N)/\mathbf{Z}/2$. Computing the differential one interprets the image of the generator of $\mathbf{Z}/2$ as $(1, \dots, 1)$. We leave this computation to the reader. QED

This computation has the merit of simplicity but does not give explicit generators of the non-trivial part of $\mathrm{Br}(X)$. One can construct quaternion algebras over the field $k(X)$ which actually belong to $\mathrm{Br}(X) \subset \mathrm{Br}(k(X))$, and whose images form a basis of $\mathrm{Br}(X)/\mathrm{Br}_0(X)$. This is another approach to the above result.

Proposition 7.1.2 *Let k be a field of characteristic* 0. *The quaternion*

algebras $A_i = (b, P_i(x))$, $i = 1, \ldots, m$, in $\mathrm{Br}(k(\mathbf{P}_k^1))$ have the property that

$$\sum_{i=1}^{m} \varepsilon_i \pi^*(A_i) \in \mathrm{Br}(k(X)), \quad \text{where} \quad \varepsilon_i \in \mathbf{Z}/2, \ i = 1, \ldots, m,$$

belongs to $\mathrm{Br}(X) \subset \mathrm{Br}(k(X))$ if and only if $\sum_{i=1}^{m} \varepsilon_i \overline{d}_i = 0 \in \mathbf{Z}/2$. Moreover, these elements generate $\mathrm{Br}(X)$ modulo $\mathrm{Br}_0(X)$.

Proof. Consider the inclusion of the generic fibre of π: $X_{k(x)} \hookrightarrow X$. The corresponding restriction map $\mathrm{Br}(X) \hookrightarrow \mathrm{Br}(X_{k(x)})$ is an embedding by a general result of Grothendieck. More precisely, $\mathrm{Br}(X)$ is the intersection of kernels of *residue* maps: $\mathrm{Br}(X_{k(x)}) \to H^1(k(C), \mathbf{Q}/\mathbf{Z})$, where C runs through all the irreducible curves contained in the fibres of π ([Grothendieck 68], Cor. 6.2).

The Brauer group of the smooth projective conic $X_{k(x)}$ is easily computed using the exact sequence (2.23). The geometric Picard group is $\mathbf{Z}$ with the trivial action of Γ_k, hence $H^1(k, Pic(X_{\overline{k(x)}})) = 0$. Since this conic has no rational point, $Pic(X_{k(x)})$ is a subgroup of $Pic(X_{\overline{k(x)}})$ of index 2. Therefore $\mathrm{Br}(X_{k(x)})$ is the quotient of the Brauer group of the base field $k(x) = k(\mathbf{P}_k^1)$ by a subgroup of order 2. On the other hand, the quaternion algebra $(b, aP(x))$ is a non-trivial element of $\mathrm{Br}(k(x))$, but its pull-back to $\mathrm{Br}(X_{k(x)})$ is trivial. We conclude that $\mathrm{Br}(X_{k(x)})$ is the quotient of $\mathrm{Br}(k(x))$ by the subgroup generated by $(b, aP(x))$.

Let $A \in \mathrm{Br}(k(x))$ be an algebra such that $\pi^*(A) \in \mathrm{Br}(X)$. We need to verify that $\pi^*(A)$ is unramified at the fibres of π. It is important to note that all the fibres are integral: they have multiplicity 1 and are irreducible (since b is not a square in the extensions of k corresponding to the irreducible factors of $P(x)$). Let P be a closed point of $\mathbf{P}_k^1$, and $res_P(A) \in H^1(k(P), \mathbf{Q}/\mathbf{Z})$ be the residue of A at P. If the fibre X_P of π at P is geometrically integral, the field $k(P)$ is integrally closed in $k(X_P)$. In this case the natural pull-back homomorphism $\pi^* : H^1(k(P), \mathbf{Q}/\mathbf{Z}) \to H^1(k(X_P), \mathbf{Q}/\mathbf{Z})$ is injective. Since $res_{X_P}(\pi^*(A)) = \pi^*(res_P(A))$, the algebra $\pi^*(A)$ is unramified at X_P if and only if A is unramified at P. By our assumption the fibre at infinity is geometrically integral, hence A is unramified at infinity. Let us now consider the opposite case. Then the integral closure of $k(P)$ in $k(X_P)$ is $k(P)(\sqrt{b})$, and the kernel of $\pi^* : H^1(k(P), \mathbf{Q}/\mathbf{Z}) \to H^1(k(X_P), \mathbf{Q}/\mathbf{Z})$ is the class of b in $H^1(k(P), \mathbf{Z}/2) = k(P)^*/k(P)^{*2} \subset H^1(k(P), \mathbf{Q}/\mathbf{Z})$. (By the assumptions at the beginning of this section this class is non-trivial.) Hence the residue of A at such a point P is either trivial or the class of b. If P is given by $P_i(x) = 0$ we set $\varepsilon_i = 0$ in the first case, and $\varepsilon_i = 1$ in the second case.

The algebra A_i is unramified on $\mathbf{P}_k^1$ except at the closed point given by $P_i(x) = 0$, where the residue is the class of b, and possibly at infinity, where the residue is the class of $b^{\bar{d}_i}$. Hence $A - \sum_{i=1}^m \varepsilon_i A_i$ is ramified at most at infinity where the residue is the class of $b^{\sum \varepsilon_i \bar{d}_i}$. This algebra belongs to $\mathrm{Br}(\mathbf{A}_k^1) = \mathrm{Br}(k)$ (this is true for any variety V such that $\bar{k}[V]^* = \bar{k}^*$ and $Pic(\overline{V}) = 0$; cf. (2.23)). Hence $A = \sum_{i=1}^m \varepsilon_i A_i + A_0$, where $A_0 \in \mathrm{Br}(k)$. We also have $\sum_{i=1}^m \varepsilon_i \bar{d}_i = 0 \in \mathbf{Z}/2$. This condition is sufficient as well as necessary for the corresponding linear combination to give rise to an element of $\mathrm{Br}(X)$. QED

7.2 Châtelet surfaces

This section is devoted to the proof of the following theorem.

Theorem 7.2.1 (Colliot-Thélène–Sansuc–Swinnerton-Dyer) *The Manin obstruction to the Hasse principle and weak approximation is the only obstruction for smooth projective models of Châtelet surfaces.*

Proof. It is enough to show that torsors of some type over such surfaces satisfy the Hasse principle and weak approximation. Considering the torsors related to the natural morphism $\pi : X \to \mathbf{P}_k^1$ sending (x, y, z) to x, we are led by Proposition 4.4.8 to the analysis of the arithmetic of the variety $Y = Y_\alpha$ given by equations (4.39). We keep the notation of Section 4.4. All we need to know about $Y \subset \mathbf{P}_k^7$ is that this complete intersection of two quadrics contains two skew $\mathbf{P}^3$'s conjugate over $k(\sqrt{b})$, and that $\overline{Y}$ can be given by

$$
\begin{aligned}
x_0 y_0 + x_1 y_1 + x_2 y_2 + x_3 y_3 &= 0, \\
\epsilon_0 x_0 y_0 + \epsilon_1 x_1 y_1 + \epsilon_2 x_2 y_2 + \epsilon_3 x_3 y_3 &= 0,
\end{aligned}
\tag{7.3}
$$

with $\epsilon_i \neq \epsilon_j$ for $i \neq j$. In this presentation the two skew $\mathbf{P}^3$'s are given by $x_i = 0$ and by $y_i = 0$, respectively. Let us call them Π_+ and Π_-. It follows from (7.3) that Y contains eight singular points given by the vanishing of all coordinates but one. The pencils of quadrics passing through Y contains precisely four quadrics of rank 6, all the other having rank 8. The following technical lemma allows us to find on Y a 'convenient' pair of skew conjugate lines. This additional structure will play a crucial rôle in the proof.

Lemma 7.2.2 *At least one of the two following statements is true:*
(a) there is a projective line $L_+ \subset \Pi_+$ defined over $k(\sqrt{b})$ such that the

span of L_+ and its conjugate $L_- \subset \Pi_-$ intersects Y in a skew quadrilateral (a cycle of four lines) contained in the smooth locus of Y;

(b) Y contains $\mathbf{P}^1_k$; in this case Y is k-rational.

Proof. Let $Z = R_{k(\sqrt{b})/k} G(1, \Pi_+)$ be the Weil descent of scalars of the Grassmannian of projective lines in Π_+. We have $\overline{Z} = G(1, \Pi_+) \times_{\overline{k}} G(1, \Pi_-)$. A k-point of Z is a pair (a line in Π_+ defined over $k(\sqrt{b})$, its conjugate line in Π_-). To a $\overline{k}$-point of Z, given by a $\overline{k}$-line $L_+ \subset \Pi_+$ and a $\overline{k}$-line $L_- \subset \Pi_-$, we associate the $\mathbf{P}^3 = \langle L_+, L_- \rangle$ spanned by them. This defines a morphism from Z to the Grassmannian of $\mathbf{P}^3$'s in the ambient $\mathbf{P}^7$. We claim that there is a dense open subset of Z such that $\langle L_+, L_- \rangle \cap Y$ is the union of L_+, L_- and some other curves; moreover, L_+ and L_- do not pass through singular points of Y.

Consider the closed subset $T \subset Y \times_k Z$ consisting of $y \in Y(\overline{k})$, $(L_+, L_-) \in Z(\overline{k})$, such that $y \in \langle L_+, L_- \rangle$. If we choose L_+ as the line given by $x_2 = x_3 = 0$, and L_- given by $y_2 = y_3 = 0$, and we see from (7.3) that the intersection of Y with their span has dimension 1. (These lines themselves do not fit because they contain singular points of Y.) The projection $T \to Z$ is proper with fibres of dimension at least 1. By upper semi-continuity of the dimension we get a dense open subset of Z where the fibres have dimension 1. We shrink this subset by requiring that the lines L_+ and L_- do not pass through the singular points of Y. Since k is infinite, this dense open set contains a k-point.

In other words, we can find $L_+ \subset \Pi_+$ defined over $k(\sqrt{b})$ and its conjugate line $L_- \subset \Pi_-$ such that these lines do not pass through the singular points of Y, and such that their span $\mathbf{P}^3_k$ intersects Y in L_+, L_- and a curve of degree 2. It is not hard to see that this residual curve cannot be a conic. (Then one of the quadrics of the pencil would contain the plane spanned by this conic, hence also another plane, and this other plane would intersect any other quadric of the pencil in a conic or a pair of intersecting lines.) The same argument shows that the residual curve cannot be a pair of intersecting lines. The remaining possibility which we want to exclude is a double line. This line must be k-rational. This case is quite possible, but then everything becomes much simpler since Y is k-rational. Indeed, a geometrically integral intersection of two quadrics with a k-line not contained in its singular locus is birationally equivalent to a variety fibred into quadrics over $\mathbf{P}^1_k$ with a section. (This is a classical observation: take a smooth point on the k-line, and cut Y by hyperplanes containing the tangent space at this point. Each such hyperplane section is birationally equivalent to a quadric, and the line defines a point in it, giving a section.) QED

In case *(b)* the theorem is obvious, so that we assume that we are in case *(a)*, and we fix L_+ and L_- from now on. Let $\Pi = \langle L_+, L_- \rangle$. We now choose a new coordinate system. Let Π be given by $X_4 = X_5 = X_6 = X_7 = 0$; these coordinates are defined over k. We can choose $\overline{k}$-coordinates in Π so that the four lines are given by $Z_i = Z_j = 0$, where $i = 0$ or $i = 1$, and $j = 2$ or $j = 3$. After modifying Z_i by an appropriate linear form in the X_j's we write the equations of Y in the following form:

$$Z_0 Z_1 + Z_2 L_2 + Z_3 L_3 + Q_1 = Z_2 Z_3 + Z_0 L_0 + Z_1 L_1 + Q_2 = 0, \qquad (7.4)$$

where L_i (resp. Q_j) are linear (resp. quadratic) forms in X_4, X_5, X_6, X_7. A straightforward calculation with the Jacobian matrix of (7.4) shows that the condition that our skew quadrilateral is contained in the smooth locus of Y is equivalent to the condition that none of the linear forms L_0, L_1, L_2, L_3 vanishes identically.

Lemma 7.2.3 *Let* $\mathbf{P}^4_{\overline{k}}$ *contain* Π. *Then* $\overline{Y} \cap \mathbf{P}^4_{\overline{k}}$ *is not integral if and only if one of the following cases occurs:* $L_2 = L_3 = Q_1 = 0$, $L_0 = L_1 = Q_2 = 0$, $L_0 L_1 = L_2 L_3 = Q_1 = Q_2 = 0$ *(identically on* $\mathbf{P}^4_{\overline{k}}$, *or, equivalently, on any point of* $\mathbf{P}^4_{\overline{k}} \setminus \Pi$).

Proof. Since $\overline{Y} \cap \mathbf{P}^4_{\overline{k}}$ is of pure dimension 2, there are three cases to consider: this is a double quadric in $\mathbf{P}^3$, the union of two such quadrics, or it contains a plane. The first case leads to a contradiction: here our surface is contained in $\mathbf{P}^3$, which must then coincide with Π, but the intersection with Π is 1-dimensional. In the second case our surface is contained in the union of two $\mathbf{P}^3$'s. This union is then a quadric of the pencil of quadrics containing the surface. It has rank 2, and can only be one of the two forms in (7.4) restricted to $\mathbf{P}^4_{\overline{k}}$. This implies that we have either $L_2 = L_3 = Q_1 = 0$ or $L_0 = L_1 = Q_2 = 0$. If our surface contains a plane, then this plane passes through one of the four lines of the quadrilateral, say through $Z_0 = Z_2 = 0$. The plane is then spanned by this line and some point $(z_0 : z_1 : z_2 : z_3 : x_4 : x_5 : x_6 : x_7)$. The points of this plane not in Π have coordinates $(z_0 : Z_1 : z_2 : Z_3 : x_4 : x_5 : x_6 : x_7)$, where Z_1 and Z_3 are indeterminates. After this substitution equations (7.4) must vanish identically. This implies the vanishing of Q_1, Q_2, L_1 and L_3 at the point $(0 : 0 : 0 : 0 : x_4 : x_5 : x_6 : x_7)$. Together with Π this point spans our $\mathbf{P}^4_{\overline{k}}$. To check that the conditions of the lemma are sufficient is straightforward. QED

Fibring Y by $\mathbf{P}^4$'s passing through Π we see that Y is birationally equivalent to a projective variety Y' equipped with a surjective morphism $f : Y' \to \mathbf{P}^3_k$ whose k-fibres are 2-dimensional intersections of two quadrics in $\mathbf{P}^4_k$ containing a pair of skew conjugate lines. At this point we make a few observations.

(1) *The generic fibre of f is smooth.* (By Bertini's theorem it is smooth away from the fixed points $\Pi \cap Y$ ([Hartshorne], III.10.9, 10.9.2). Since none of L_0, L_1, L_2, L_3 vanishes identically, the generic fibre is also smooth at the points of our skew quadrilateral – this is proved by the direct calculation with the Jacobian mentioned before Lemma 7.2.3.)

(2) *There is a closed subset $V \subset \mathbf{P}^3_k$ of dimension at most 1 such that the restriction of f to $U = \mathbf{P}^3_k \setminus V$ has geometrically integral fibres.* (This follows from the previous lemma and the fact that the L_i's do not vanish identically.)

(3) *Smooth intersections of two quadrics in $\mathbf{P}^4_k$ containing a pair of skew conjugate lines satisfy the Hasse principle and weak approximation.*

To prove (3) we note that a smooth intersection of two quadrics in $\mathbf{P}^4_k$ is a Del Pezzo surface of degree 4 (a $(\overline{k}/k)$-form of $\mathbf{P}^2$ with five points in general position blown up; 'the general position' here means that no three points are on a line, and all five points are not on a conic). Therefore, blowing down a pair of skew conjugate lines defines a birational morphism to a Del Pezzo surface of degree 6 (a $(\overline{k}/k)$-form of $\mathbf{P}^2$ with three non-collinear points blown up). Then the Hasse principle and weak approximation hold by Corollary 6.3.3.

It remains to show how the theorem follows from (1), (2) and (3).

In the course of the proof we shall use the following theorem of Lang and Weil [LW]. Let $\mathbf{F}_q$ be a finite field with q elements, and let X be a geometrically integral subvariety of $\mathbf{P}^n_{\mathbf{F}_q}$ of dimension r and degree d. Then there exists a positive constant $C(n, r, d)$ such that the number of $\mathbf{F}_q$-points on X satisfies the following inequality:

$$|\#X(\mathbf{F}_q) - q^r| < C(n, r, d)q^{r - \frac{1}{2}}.$$

One can apply these estimates to reductions of a geometrically integral variety defined over a number field. They imply that provided q is big enough so that the reduction is also geometrically integral, we can always find an $\mathbf{F}_q$-point in the reduction (even in the reduction of a given dense open subset) of our variety. Since the constant depends only on n, r and d the estimates can also be applied to families of closed subvarieties of $\mathbf{P}^n$

of the same degree and dimension. (Such are flat families; in a flat family of projective varieties the Hilbert polynomial is constant, hence so are the dimension and the degree; see [Hartshorne], III, 9.9, 9.10).

We have to find a smooth k-point on Y' which is close to a given finite collection of smooth local points $Q_v \in Y'(k_v)$, $v \in \Sigma$. In the course of the proof we can enlarge Σ, and move Q_v in a small neighbourhood in the corresponding local topology. For example, we can assume that Q_v do not belong to some Zariski closed subset.

Choose an auxiliary k-point $P \in U(k)$ such that the fibre $Y_P = f^{-1}(P)$ is smooth. Every geometrically integral smooth k-variety X has k_v-points for all but finitely many places of v. (For almost all primes of k the reduction of X is a well defined geometrically integral smooth variety, thus we can apply the Lang–Weil estimates to X to conclude that for almost all primes the reduction of X has a smooth point over the residue field. To such a point we apply Hensel's lemma to get a local point.) Now we enlarge Σ and our collection of local points $Q_v \in Y'_{smooth}(k_v)$, $v \in \Sigma$, by adding local points for all the places v such that $Y_P(k_v) = \emptyset$. Using weak approximation in $\mathbf{P}^3_k$ we find a k-point R which is close to $f(Q_v)$, $v \in \Sigma$, and such that the line $PR \simeq \mathbf{P}^1_k$ does not meet V, and such that the generic fibre of the restriction of f to PR is smooth. This is possible since $dim(V) \leq 1$, whereas the space of lines in $\mathbf{P}^3_k$ passing through P is isomorphic to $\mathbf{P}^2_k$. We note that $H = f^{-1}(PR)$ has smooth local points for all places of k. It is enough now to find a k-point in H close to a given finite collection of k_v-points for $v \in \Sigma$. Note that all the fibres of $H \to \mathbf{P}^1_k$ are geometrically integral, and only finitely many fibres are singular. Comparing the Lang–Weil estimates for the fibres and their singular loci we find a finite set of primes with the property that if the reduction of a fibre of $H \to \mathbf{P}^1_k$ at any other prime is geometrically integral, then it has a smooth point over the residue field. (See [S90b] for a general set-up of this argument.)

Enlarging Σ again we can find a model $\mathcal{H} \to \mathbf{P}^1_{\mathcal{O}_{k,\Sigma}}$ of $H \to \mathbf{P}^1_k$ with all the closed geometric fibres being geometrically integral. (Indeed, the subscheme of $\mathbf{P}^1_{\mathcal{O}_{k,\Sigma}}$ corresponding to geometrically reducible or non-reduced fibres does not intersect $\mathbf{P}^1_k$ – the generic fibre of the structure morphism to $Spec(\mathcal{O}_{k,\Sigma})$ – hence is contained in the union of finitely many closed fibres of $\mathbf{P}^1_{\mathcal{O}_{k,\Sigma}} \to Spec(\mathcal{O}_{k,\Sigma})$. We only have to include the corresponding primes in Σ.)

Using weak approximation on the base $PR \simeq \mathbf{P}^1_k$ we find a k-point M on this line such that Y_M is smooth and has k_v-points close to the given k_v-points for $v \in \Sigma$. Note that we included in Σ all 'small' primes, so that now the reduction of Y_M at any place not in Σ is geometrically integral and has a smooth point over the residue field. Using Hensel's lemma we can lift it to a smooth k_v-point of Y_M, $v \notin \Sigma$. Summing up, we have a k-fibre with points everywhere locally. By (3) it has a k-point which is close to our initial collection of local points on Y'. This finishes the proof of the theorem. QED

The knowledge that a class of varieties satisfies the Hasse principle is more practical than the knowledge that the Manin obstruction to it is the only one. In this respect we have the following statement which is an application of Proposition 7.1.1.

Corollary 7.2.4 *The class of Châtelet surfaces given by the equation $y^2 - bz^2 = aP(x)$, where $P(x)$ is irreducible of degree 4, satisfies the Hasse principle and weak approximation. Surfaces given by $y^2 - bz^2 = aP(x)$, where $P(x)$ is irreducible of degree 3, satisfy weak approximation.*

Let k be an arbitrary field. Let us show that an X-torsor of the type considered above with a k-point is k-rational. Indeed, by Proposition 4.4.8 such a torsor is birationally equivalent to the product of $\mathbf{A}_k^1$, the conic $y^2 - bz^2 = aN(\alpha)$, and Y_α given by equations (4.39). If the torsor corresponding to some α contains a k-point it is birationally equivalent to the product of $\mathbf{A}_k^2$ and $Y = Y_\alpha$. Let $P \in Y(k)$. The variety Y is an intersection of two quadrics in $\mathbf{P}_k^7$ containing a pair of skew conjugate $\mathbf{P}^3$'s defined over $k(\sqrt{b})$; we called them Π_+ and Π_-. Let $P_+ = \langle P, \Pi_+ \rangle \cap \Pi_-$, and $P_- = \langle P, \Pi_- \rangle \cap \Pi_+$, where $\langle \cdot, \cdot \rangle$ denotes the linear span. Consider the line L passing through collinear points P, P_+ and P_-. Every quadric of the pencil of quadrics containing Y contains these three points, hence contains L. By the argument at the end of the proof of Lemma 7.2.2, any intersection of two quadrics with a k-line not entirely contained in the singular locus is k-rational. But $dim(Y_{sing}) = 0$, hence our statement is proved.

The effect of the rationality of torsors is that when k is a number field, only finitely many torsors of a given type contain k-points, thus the set $X(k)$ is represented as a disjoint union of $f_\alpha(Z_\alpha(k))$, where $f_\alpha : Z_\alpha \to X$ is a finite family of dominant maps of k-rational varieties Z_α. Such a description was first found by F. Châtelet in the case when $P(x)$ is a product of four linear factors [Châtelet]; see [Manin, CF], VI. In other words, the set of k-points of X consists of finitely many classes, and each class is parametrized by rational functions in finitely many variables. As Manin proves in *op. cit.* two variables are not enough.

Another application of this technique is a negative solution [BCSS] of the Zariski conjecture: must a stably rational variety always be rational? Consider a Châtelet surface X over a field k of characteristic different from 2, given by $y^2 - bz^2 = P(x)$, where $P(x)$ is a separable irreducible polynomial of degree 3 with discriminant $b \notin k^{*2}$. Then $X \times_k \mathbf{A}_k^3$ is birationally equivalent to $\mathbf{A}_k^5$, but X is not k-rational. The idea consists of proving the k-rationality of a certain X-torsor (which is isomorphic to a variety Y_α for some α) under a k-*rational* torus. We refer the reader to the original paper [BCSS] for details, as well as for a negative solution of the same problem over an algebraically closed field (which builds on this example with $k = \mathbf{C}(t)$).

Counter-examples to the Hasse principle.

We now explore a family of counter-examples to the Hasse principle which are variations of Iskovskih's counter-example [Iskovskih]. It is taken from ([Sansuc 82], Sect. 2) which elaborates on [CCS]. Note that in most 'real life' cases the computation of the Manin obstruction is more difficult than in this example, chosen for its particular simplicity (cf. [CKS] or [C86] in the case of diagonal cubic surfaces).

Consider the Châtelet surface X_c over $k = \mathbf{Q}$ given by

$$y^2 + 3z^2 = (c - x^2)(x^2 - c + 1) \tag{7.5}$$

where $c \in \mathbf{Z}$, $c \neq 0$, $c \neq 1$. One sees immediately that $X_c(\mathbb{R}) \neq \emptyset$ if and only if $c > 1$. The local solubility of (7.5) in $\mathbf{Q}_v$ for any finite place v imposes no restriction on c. This is easily seen for $p \neq 3$ by setting $x = p^{-1}$ and using the fact that a unit is a norm for an unramified extension. For $p = 3$ the solubility of (7.5) is established by a case by case computation.

Consider the quaternion algebra $A = (c - x^2, -3)$ over $k(X_c)$. By Proposition 7.1.2 the image of its class generates $\mathrm{Br}(X_c)/\mathrm{Br}_0(X_c)$, so to compute the Manin obstruction we only need to compute the sum $\sum_{v \in \Omega} \mathrm{inv}_v(A(P_v))$, $P_v \in X_c(\mathbf{Q}_v)$.

Statement 1: If $v \neq 3$, then $\mathrm{inv}_v(A(P_v)) = 0$ for any point $P_v \in X_c(\mathbf{Q}_v)$. This value is locally constant in the v-adic topology, hence we may assume that P_v is not contained in the fibre at infinity or in any of the sigular fibres, that is, $(c - x^2)(x^2 - c + 1) \neq 0$. We must prove that $c - x^2$ is locally a norm for the extension $\mathbf{Q}(\sqrt{-3})/\mathbf{Q}$. For $\mathbf{Q}_v = \mathbb{R}$ it is enough to note that $c - X^2 > 0$. For a finite $v \neq 3$ we only have to consider the case when p is inert for $\mathbf{Q}(\sqrt{-3})/\mathbf{Q}$. We have two possibilities: $v(x) < 0$ and $v(x) \geq 0$. In the first case $v(c - x^2)$ is even, hence this is the product of a unit, which is a norm for the unramified extension $\mathbf{Q}_v(\sqrt{-3})$, and an even power of a uniformizer, which is trivially a norm for any quadratic extension. Since $(c - x^2) + (x^2 - c + 1) = 1$, in the second case either $v(c - x^2) = 0$, in which case $c - x^2$ is a norm, or $v(x^2 - c + 1) = 0$. In the latter case from the equation of X_c it follows that $c - x^2$ is a norm multiplied by a unit, hence is a norm.

Statement 2: For $v = 3$ and $c = 3^{2n+1}(3m + 2)$ we have $\mathrm{inv}_3(A(P_3)) = \frac{1}{2}$ *for any point $P_3 \in X_c(\mathbf{Q}_3)$, whereas for other values of c the local invariant takes both values 0 and $\frac{1}{2}$.* This purely local computation is omitted here.

Conclusion. When the sum of local invariants is never 0, the Manin obstruction tells us that no $\mathbf{Q}$-point can exist on X_c. This happens for $c = 3^{2n+1}(3m + 2)$, whereas X_c has adelic points for any $c > 1$. Theorem 7.2.1 implies that in all the other cases for $c \in \mathbf{Z}$, $c > 1$, the surface X_c contains a $\mathbf{Q}$-point. (Note that in the particular case when $P(x)$ is a product of two irreducible quadratic polynomials this theorem was proved in [CCS] using an entirely different technique.)

7.3 Some intersections of two quadrics in $\mathbf{P}_k^5$

Let K/k be a quadratic extension, U_1, V_1 be K-variables, U_2, V_2 be their conjugates over k, φ_1 and φ_2 be quadratic forms of rank 2 with coefficients in K, which are conjugate over k. Let us consider the projective k-variety in $\mathbf{P}_k^5$ defined by the following equations:

$$\begin{aligned} U_1^2 - bV_1^2 &= \varphi_1(Z, T), \\ U_2^2 - bV_2^2 &= \varphi_2(Z, T). \end{aligned} \tag{7.6}$$

Theorem 7.3.1 (Colliot-Thélène–Sansuc–Swinnerton-Dyer) *Suppose that φ_1 and φ_2 have no common factor over $\overline{k}$. Then the Manin obstructions to the Hasse principle and weak approximation are the only obstructions for smooth proper models of k-varieties given by (7.6).*

Proof. Let us note that each equation (7.6) is a quadric in four variables, and the projection $(U_i : V_i : Z : T) \mapsto (Z : T)$ modifies it birationally into a smooth and proper surface which is a conic bundle over $\mathbf{P}_K^1$ with two singular fibres. Let us denote it by $S_i \to \mathbf{P}_K^1$. This surface can be given by the equation

$$U_i^2 - bV_i^2 = \varphi_i(x, 1)W_i^2, \ i = 1, 2, \tag{7.7}$$

homogeneous in U_i, V_i, W_i, exactly in the same way as in Section 7.1. Then a smooth and proper model X of the k-variety given by (7.6) is the quotient scheme $S_1 \times_{\mathbf{P}_K^1} S_2$ by the natural action of $Gal(K/k)$ which swaps S_1 and S_2. The set of $\overline{k}$-points of X is the fibre product of $S_1(\overline{k})$ and $S_2(\overline{k})$ over $\mathbf{P}_{\overline{k}}^1(\overline{k})$. The k-variety X is proper. The $\overline{k}$-points of $\mathbf{P}_k^1$ corresponding to singular fibres of $S_1 \to \mathbf{P}_K^1$ and $S_2 \to \mathbf{P}_K^1$ are disjoint by our assumption. A simple local calculation shows that X is smooth. The singular fibres of $f : X \to \mathbf{P}_k^1$ are located over the zeros of the polynomial $P(x) = \varphi_1(x,1)\varphi_2(x,1) \in k[x]$, where $x = Z/T$.

According to Corollary 4.4.6 any X-torsor associated with the morphism $f : X \to \mathbf{P}_k^1$ is birationally equivalent to the fibre product of X and some $W = W_\alpha$ over $\mathbf{P}_k^1$, where W is given by equations (4.34). One sees immediately that W is birationally equivalent to the affine cone over the variety Y studied in the proof of the previous theorem. In particular, W satisfies the Hasse principle and weak approximation. On the other hand, looking at equations (4.39) one easily finds a birational transformation (similar to the one in the proof of Proposition 4.4.8) showing that $X \times_{\mathbf{P}_k^1} W$ is birationally equivalent to the product of W and the Weil descent $R_{K/k}(C)$ of a smooth, projective K-conic. By the fundamental property of the Weil descent we have

$$R_{K/k}(C)(k_v) = C(K \otimes_k k_v) = \prod_{w|v} C(K_w)$$

where w runs over the places of K over v. Hence

$$R_{K/k}(C)(k) = C(K) \subset C(\mathbb{A}_K) = R_{K/k}(C)(\mathbb{A}_k)$$

is a dense subset. This means that $R_{K/k}(C)$ satisfies the Hasse principle and weak approximation. We have shown that the class of X-torsors associated with the morphism $f : X \to \mathbf{P}_k^1$ satisfies the Hasse principle and weak approxmation. By Corollary 6.1.3 this implies the statement of the theorem. QED

Corollary 7.3.2 *If, in the assumptions of the previous theorem, the polynomial* $P(x) = \varphi_1(x,1)\varphi_2(x,1) \in k[x]$ *is irreducible, then the k-varieties given by (7.6) satisfy the Hasse principle and weak approximation.*

Proof. We need to compute $H^1(k, Pic(\overline{X}))$. This computation is completely analogous to Proposition 7.1.1. The Galois group acts on the irreducible components of the fibres of f in the same way as on the components of the conic bundles considered in Section 7.1. In particular, the Γ_k-submodule of $Pic(\overline{X})$ generated by these components is isomorphic to

N in the notation of that section, and we already know $H^1(k, N)$. In our assumptions this group is $\mathbf{Z}/2$. Denoting by $X_{\overline{k}(x)}$ the generic fibre of $\overline{X} \to \mathbf{P}_{\overline{k}}^1$ we need to compute the cokernel of $Pic(\overline{X})^{\Gamma_k} \to Pic(X_{\overline{k}(x)})^{\Gamma_k}$. Now $Pic(X_{\overline{k}(x)}) = \mathbf{Z} \oplus \mathbf{Z}$ is generated by the images of the classes of divisors $l_1 \times_{\mathbf{P}_{\overline{k}}^1} S_2$ and $S_1 \times_{\mathbf{P}_{\overline{k}}^1} l_2$, where $l_i \subset S_i$ is given by $U_i - \sqrt{b}V_i = W_i = 0$, $i = 1, 2$, in the coordinates of (7.7). It follows that $Pic(X_{\overline{k}(x)})^{\Gamma_k}$ is generated by the image of the sum of these two classes. On the other hand, $Pic(\overline{X})^{\Gamma_k}$ is generated by certain sums of components of the fibres and the sum of these two classes plus the sum of two similar classes obtained by changing the sign of $\sqrt{b}$. It follows that the cokernel is $\mathbf{Z}/2$, and we conclude that $H^1(k, Pic(\overline{X})) = 0$. QED

7.4 Conic bundles with six singular fibres

In this section we prove the following theorem.

Theorem 7.4.1 (Swinnerton-Dyer) *The Manin obstruction to the Hasse principle and weak approximation is the only obstruction for smooth projective surfaces over a number field k which are birationally equivalent to the surface*

$$y^2 - bz^2 = af(x)g(x), \qquad (7.8)$$

where $f(x)$ and $g(x)$ are monic irreducible polynomials over k of degrees $deg(f(x)) = 2$ and $deg(g(x)) = 4$, $a \in k^*$.

The theorem remains true without the irreducibility assumption [SD99]. (This paper only treats the Hasse principle but not weak approximation.) A unirationality argument at the end of the proof in [SD99] contains a difficulty, which can be overcome but the proof becomes more involved and is not given here.

Proof of theorem. It is not surprising that this proof relies on the thorough analysis of the case of four singular fibres. We keep the same notation as in the proof of Theorem 7.2.1. As in that proof, our goal is to show that torsors related to the natural morphism to $\mathbf{P}_k^1$ sending (x, y, z) to x satisfy the Hasse principle and weak approximation. By Proposition 4.4.8 it is enough to prove these properties for the descent variety $Y = Y_\alpha \subset \mathbf{P}_k^{11}$ given by equations (4.39):

$$\alpha_i(u - e_i v) = U_i^2 - bV_i^2, \quad i = 1, \dots, 6. \qquad (7.9)$$

Here u, v are variables with values in k, and U_i, V_i are variables with values in $k(e_i)$, such that U_i and U_j are conjugate when e_i and e_j are conjugate (and similarly for V_i and V_j). The constants $\alpha_i \in k(e_i)^*$ are such that α_i and α_j are conjugate when e_i and e_j are conjugate. We arrange that e_1, e_2, e_3, e_4 are the roots of $g(x) = 0$, and e_5 and e_6 are the roots of $f(x) = 0$. Multiplying c by a square in k we can assume that c is in $\mathcal{O}_k$, the ring of integers of k. We assume that c is not a square in any $k(e_i)$, $i = 1, \dots, 6$ (see the beginning of Section 7.1).

We shall work over $\overline{k}$ respecting conjugacy, so that the varieties and the morphisms that we shall construct will be actually defined over k. Set

$$x_i = \alpha_i^{-1}(U_i - \sqrt{b}V_i), \quad y_i = U_i + \sqrt{b}V_i, \quad i = 1, \dots, 6.$$

Over $\overline{k}$ equations (4.39) can be rewritten as

$$u - e_i v = x_i y_i, \quad i = 1, \dots, 6. \tag{7.10}$$

Eliminating u and v from (7.10) we see that Y is a complete intersection of four quadrics in $\mathbf{P}_k^{11}$, and that no quadric passing through Y has rank less than 6. Let us call $X \subset \mathbf{P}_k^7$ the descent variety for the Châtelet surface $y^2 - bz^2 = g(x)$ obtained from the equations (4.39) defining Y by considering only the equations corresponding to $i = 1, 2, 3, 4$. This variety was the main object of study in the previous section. The equations of $\overline{X}$ can be written as

$$x_3 y_3 = \frac{e_2 - e_3}{e_2 - e_1}x_1 y_1 - \frac{e_1 - e_3}{e_2 - e_1}x_2 y_2, \quad x_4 y_4 = \frac{e_2 - e_4}{e_2 - e_1}x_1 y_1 - \frac{e_1 - e_4}{e_2 - e_1}x_2 y_2.$$

$$\tag{7.11}$$

Let $W \subset X$ be the complement of the union of $\mathbf{P}^3$'s given by the condition that for every $i = 1, 2, 3, 4$ either $x_i = 0$ or $y_i = 0$. (In Section 4.4, Example 3, we called W_α the affine cone over this variety.) In terms of (7.10) the open subset $W \subset X$ is given by $(u, v) \neq (0, 0)$. One can also characterize W by the property that at most one monomial $x_i y_i$ equals 0, for $i = 1, 2, 3, 4$. The equations of $\overline{Y}$ are obtained by adding to (7.11) the following two equations:

$$x_5 y_5 = \frac{e_2 - e_5}{e_2 - e_1}x_1 y_1 - \frac{e_1 - e_5}{e_2 - e_1}x_2 y_2, \quad x_6 y_6 = \frac{e_2 - e_6}{e_2 - e_1}x_1 y_1 - \frac{e_1 - e_6}{e_2 - e_1}x_2 y_2.$$

$$\tag{7.12}$$

Forgetting the coordinates x_5, y_5, x_6, y_6 defines a surjective projective morphism $\pi : Y' \to X$ where Y' is a projective k-variety birationally equivalent to Y. Let $F = (u - e_5 v)(u - e_6 v)$ be the product of the right hand sides of (7.12). Let $W' \subset W$ be the dense open subset given by $F \neq 0$. The restriction of π to W' has smooth $\bar{k}$-fibres which are products of two conics.

By the proof of Theorem 7.2.1 the variety X which contains points everywhere locally also contains a global point, and these are dense in the set $X(k_S)$ for any finite set of places $S \subset \Omega$. Such a point P is given by coordinates $u, v \in k$ and $U_i, V_i \in k(e_i)$ with the standard conjugacy assumption, satisfying the equations

$$\alpha_i(u - e_i v) = U_i^2 - bV_i^2, \quad i = 1, 2, 3, 4. \tag{7.13}$$

The constants α_i can be modified by a square in $k(e_i)$ subject to preserving conjugacy. The following auxiliary lemma shows that we can modify α_i's by conjugate squares so that (7.13) will have the following property: if (7.13) has a solution with u and v in $\mathcal{O}_k$, then for the same u and v it has a solution with $U_i, V_i \in \mathcal{O}_{k(e_i)}$. This will be used later in the proof when we shall consider the reduction of P modulo primes of k.

Lemma 7.4.2 *For $i = 1, 2, 3, 4$ there exists a constant $A_i = A_i(b) \in \mathcal{O}_{k(e_i)}$ such that if $\alpha_i \in \mathcal{O}_{k(e_i)}$ is an integer divisible by A_i^2, then the following property holds: given a solution (u, v, U_i, V_i) of (7.13) with $u, v \in \mathcal{O}_k$, there exists a solution (u, v, U_i', V_i') with U_i' and V_i' in $\mathcal{O}_{k(e_i)}$.*

Proof. It is enough to prove the following statement. Let K be a number field, c an integer in K but not a square, $L = K(\sqrt{c})$. Let $I_1, \ldots, I_n$ be a set of representatives of the ideal classes of L, and let $A = A(K, c) \in \mathcal{O}_K$ be a constant divisible by the ideal $2cN_{L/K}(I_r)$ for every $r = 1, \ldots, n$. We claim that if $D \in \mathcal{O}_K$ is of the form $U^2 - cV^2$, for some $U, V \in K$, and if $A^2|D$, then $D = U^2 - cV^2$ for some $U, V \in \mathcal{O}_K$.

Let σ be the generator of $Gal(L/K)$. Let $D = dA^2$, $d \in \mathcal{O}_K$. We can represent d as the norm of $u + \sqrt{c}v$ with $u, v \in K$. We can write $(u + \sqrt{c}v) = IJ^{-1}$, where I and J are coprime ideals of L. Let r, $1 \leq r \leq n$, be such that I_rJ is a principal ideal, $I_rJ = (B)$ for some $B \in L$. Then $I\sigma(I) = (d)J\sigma(J)$, hence $\sigma(J)|I$. The norm of $B\sigma(B)^{-1}(u + \sqrt{c}v)$ is d. On the other hand, the denominator of

$$(B\sigma(B)^{-1}(u + \sqrt{c}v)) = I_rJ\sigma(I_r)^{-1}\sigma(J)^{-1}IJ^{-1}$$

divides $\sigma(I_r)$. Multiplying this element of L by A we get an element $U + \sqrt{c}V$ with $U, V \in \mathcal{O}_K$, whose norm is D. QED

Let $P \in X_{smooth}(k)$, and let T_P be the tangent space to X at P. Since the varieties which we consider are intersections of quadrics, $X \cap T_P$ is the affine cone over an intersection of two quadrics in $\mathbf{P}_k^4$ which we denote by

X_P. In other words, $X \cap T_P$ consists of projective lines passing through P and contained in X. Let T be the k-variety parametrizing pairs (P, L), where P is a point of X, and L is a projective line passing through P and contained in X. We denote by $\tau : T \rightarrow X$ the projection $(P, L) \mapsto P$. Recall that $\Pi_+ \subset X$ is given by $x_1 = x_2 = x_3 = x_4 = 0$, and $\Pi_- \subset X$ by $y_1 = y_2 = y_3 = y_4 = 0$.

Lemma 7.4.3 *The restriction of the natural projection $\tau : T \rightarrow X$ to the open set $X \setminus (\Pi_+ \cup \Pi_-)$ has a section $P \mapsto L_P$ given by the line*

$$L_P = \langle \langle P, \Pi_+ \rangle \cap \Pi_-, \langle P, \Pi_- \rangle \cap \Pi_+ \rangle$$

passing through P, and entirely contained in X. Let Q_P be the point on X_P defined by L_P. Then Q_P is singular on X_P if and only if P has at least five zero coordinates x_i, y_j, and for every $i = 1, 2, 3, 4$ either $x_i = 0$ or $y_i = 0$.

Proof. Recall that X is an intersection of quadrics containing three distinct points of L_P, that is, P, $L_P \cap \Pi_+$, $L_P \cap \Pi_-$; then every quadric in the pencil of quadrics defining X contains L_P, hence so does X. It is clear that Q_P is smooth on X_P if and only if some point of L_P different from P is smooth on $X \cap T_P$. Take any point P' on L_P different from the above three points; then P' is smooth in $X \cap T_P$ provided that P is in W. This can be easily seen from the Jacobian matrix of $X \cap T_P$, using the fact that at most one monomial $x_i y_i$ vanishes when P is in W. If P is not in W, then all monomials $x_i y_i$ vanish. In other words, for every $i = 1, 2, 3, 4$ either $x_i = 0$ or $y_i = 0$. Thus the Jacobian matrix has at most four non-zero columns. Computing the determinant of the resulting 4×4 matrix for a sufficiently general P' in L_P we see that it equals a non-zero constant multiplied by the product of the remaining coordinates of P. This proves the lemma. QED

By this lemma if Q_P is singular then at least one of the following conditions holds:

(a) $P \in \Pi_+ \cup \Pi_-$,

(b) $x_i y_j y_k \neq 0$ or $x_i x_j y_k \neq 0$ for some pairwise different i, j, k, and all the remaining coordinates equal zero,

(c) there exist $i \neq j$ such that $x_i = y_i = x_j = y_j = 0$, $i \neq j$. This is equivalent to $U_i = V_i = U_j = V_j = 0$.

This lemma means that X has 'many' k-lines, through every k-point of X passes at least one k-line. Our main goal is to find a k-line $L \subset X$ such that $H = \pi^{-1}(L) \subset Y$ has points everywhere locally, and such that for a prescribed finite set of places, H has local points close to a given collection of local points on Y. If, moreover, H satisfies the Hasse principle and weak approximation, the theorem will be proved.

The following lemma explains how the structure of X_P depends on P.

Lemma 7.4.4 *Let $P \in W(k)$, and p_i be the product of coordinates x_i and y_i of P, $i = 1, 2, 3, 4$. Then if no p_i vanishes, X_P is smooth. In the opposite case, X_P is a geometrically integral intersection of two quadrics in $\mathbf{P}^4$ with two isolated singular points.*

Recall that if a point P is in W, then at most one p_i can vanish.

Proof of lemma. We can work over $\overline{k}$. By symmetry we assume that $p_3 p_4 \neq 0$. Adding to (7.11) the equations of the tangent hyperplanes to both quadrics at P, we eliminate y_3 and y_4 from the resulting system. Considering the base of the cone $X \cap T_P$ amounts to the same as cutting $X \cap T_P$ by a hyperplane not passing through P, say the one given by $x_3 = 0$. Making this substitution we get two quadratic forms Φ_0 and Φ_1 in five variables (we can arrange that Φ_0 is the sum of two hyperbolic planes). A straightforward computation shows that $det(\Phi_1 - t\Phi_0) = 0$ has four roots 0, α, β, and $\alpha\beta p_3/p_4$, where $\alpha = (e_2 - e_4)(e_2 - e_3)^{-1}$, $\beta = (e_1 - e_4)(e_1 - e_3)^{-1}$. It is a well known and easily verified fact that the variety given by $\Phi_0 = \Phi_1 = 0$ in $\mathbf{P}^4$ is smooth if and only if there are precisely five degenerate quadrics in the pencil $\lambda\Phi_0 + \mu\Phi_1$. (Indeed, over an algebraically closed field two quadratic forms, one of which is non-degenerate, can be made simultaneously diagonal. If one of them is represented by the identity matrix, and the other by $diag(\lambda_1, \ldots, \lambda_n)$, then the variety defined by

$$\sum z_i^2 = \sum \lambda_i z_i^2 = 0$$

is smooth if and only if $\lambda_i \neq \lambda_j$ for $i \neq j$. If $\lambda_i = \lambda_j$ for only one pair (i, j), then one finds exactly two singular points on it; these are given by $z_i^2 + z_j^2 = 0$, and $z_k = 0$ for $k \neq i, j$.) In our case the degenerate quadrics correspond to the roots of $det(\Phi_1 - t\Phi_0) = 0$ and $t = \infty$. Hence X_P is smooth unless $\alpha\beta p_3/p_4$ coincides with α or β. The first case implies $p_1 = 0$, and the second one implies $p_2 = 0$, as is immediately seen from (7.11). In any of these cases we are led to a geometrically integral intersection of two quadrics with two singular points. QED

Lemma 7.4.5 *Let $P \in X(k)$ be such that all its coordinates x_i and y_j are non-zero. Then X_P satisfies weak approximation (and hence so does $T_P \cap X$).*

Proof. We know that X_P is a smooth intersection of two quadrics in $\mathbf{P}_k^4$. Thus X is a Del Pezzo surface of degree 4. Such a surface contains 16 lines defined over $\overline{k}$ ([Manin, CF], IV). These lines can be identified as follows. Consider the projective 3-dimensional space Π_+ (resp. Π_-) given by $x_i = 0$, $i = 1, 2, 3, 4$ (resp. $y_i = 0$, $i = 1, 2, 3, 4$). The image l_+ of $\Pi_+ \cap T_P$ in X_P, and the image l_- of $\Pi_- \cap T_P$ in X_P, are projective lines which meet at Q_P, and are globally defined over k. Let $\Pi_+^{(i)}$ be the projective 3-dimensional space given by $x_j = 0$, $j \in \{1, 2, 3, 4\} \setminus \{i\}$, $y_i = 0$. We define $\Pi_-^{(i)}$ by exchanging the rôles of x_j's and y_i's. Let $l_+^{(i)}$ be the image of $\Pi_+^{(i)} \cap T_P$ in X_P, and define $l_-^{(i)}$ similarly. The intersection indices are as follows:

$$(l_+^{(i)}, l_-^{(j)}) = \delta_{ij}, \ (l_+, l_+^{(i)}) = 1, \ (l_+, l_-^{(i)}) = 0, \ (l_-, l_+^{(i)}) = 0, \ (l_-, l_-^{(i)}) = 1.$$

It follows from the irreducibility of $g(x)$ that the $l_+^{(i)}$'s and the $l_-^{(i)}$'s form one orbit of Γ_k acting on the set of 16 lines. It follows from the intersection indices that for any i the sum of l_+, l_-, $l_+^{(i)}$, $l_-^{(i)}$ is a quadrilateral spanning a $\mathbf{P}^3$, that is, is a hyperplane section of X_P. Hence the classes of $l_+^{(i)} + l_-^{(i)}$ in $Pic(\overline{X_P})$ are all equal, and define a conic bundle structure $X_P \to \mathbf{P}_k^1$. (Cutting X_P by $\mathbf{P}^3$'s passing through the plane spanned by l_+ and l_-.) Both l_+ and l_- are sections of this morphism. Let $E = f_*(\mathcal{O}(l_+ + l_-))$ be the direct image of the invertible sheaf $\mathcal{O}(l_+ + l_-)$ which is relatively ample over $\mathbf{P}_k^1$. Then we get an embedding $X_P \hookrightarrow \mathbf{P}(E^*)$. Its restriction to $\mathbf{A}_k^1$ complementary to the fibre containing Q_P is an embedding into $\mathbf{A}_k^1 \times_k \mathbf{P}_k^2$. It is easy to choose the coordinates so that the image is given by the equation $y^2 - bz^2 = Q(x)u^2$. Hence X_P is birationally equivalent to a Châtelet surface. The polynomial $Q(x)$ whose zeros correspond to the singular fibres is irreducible of degree 4 with residue field $k[x]/(g(x))$. By Corollary 7.2.4 such a Châtelet surface satisfies weak approximation, hence so does X_P. QED

Let Σ_0 be a finite set of places of k consisting of all the archimedean places and all the primes dividing $2b$, the α_i's, the e_i's, and the differences $e_i - e_j$, and such that $k(e_1, e_2, e_3, e_4, e_5, e_6)$ is unramified outside Σ_0, and $\mathcal{O}_{k, \Sigma_0}$ is principal.

Let m_v be the prime ideal of $\mathcal{O}_k$ corresponding to a place $v \notin \Sigma_0$. We shall denote by a tilde the reduction modulo m_v. The vanishing of F defines a closed subset of $T_P \cap X$. Let us show that it has codimension 1, and the same remains true for its reduction at any prime not in Σ_0, provided that at most one $p_i = x_i y_i \equiv 0 \bmod m_v$. Without loss of generality we assume that $p_3 = x_3 y_3 \not\equiv 0 \bmod m_v$ and $p_4 = x_4 y_4 \not\equiv 0 \bmod m_v$. Then the projection to coordinates x_1, y_1, x_2, y_2 defines a dominant, generically finite map from the reduction of $T_P \cap X$ to $\mathbf{P}^3_{\mathbf{F}_v}$, as one can easily see from equations (7.11). On the other hand, $\tilde{F}$ is a non-zero polynomial in x_1, y_1, x_2, y_2 by our choice of Σ_0. The vanishing of F thus defines a divisor on $T_P \cap X$, whose reduction is still a divisor for all primes not in Σ_0 (provided at most one p_i is congruent to 0). By the Lang–Weil estimates, there exists a constant C such that for all primes of k with norm greater than C there is a smooth $\mathbf{F}_v$-point Q^v on the reduction of $T_P \cap X$ such that $\tilde{F}(Q^v) \neq 0$.

Let Σ be the union of Σ_0 and all the primes of k with norm less than C. For all $v \notin \Sigma$ the reduction $\tilde{X}$ is a geometrically integral intersection of two quadrics in $\mathbf{P}^7_{\mathbf{F}_v}$; and similarly for Y. If $\tilde{P}$ is in $\tilde{W}$, then the reduction of X_P is a geometrically integral surface (Lemma 7.4.4), and the reduction of Q_P is a smooth point on it (Lemma 7.4.3).

Now using weak approximation on X we choose a point in $W'(k)$ in the image $\pi(Y(k_v))$ for $v \in \Sigma$, such that all its coordinates x_i and y_j are non-zero. By multiplying the coordinates U_i and V_i by a common element $\mu \in k$, and multiplying u and v by μ^2 we ensure that u and v are in $\mathcal{O}_{k,\Sigma}$, and that the ideal (u, v) is not divisible by the square of any prime not in Σ. Then we choose another point $P \in W'(k)$ with the same coordinates u and v but with $U_i, V_i \in \mathcal{O}_{k(e_i)}$, which is possible by Lemma 7.4.2. Note that since u and v are not changed, $P \in \pi(Y(k_v))$ for $v \in \Sigma$. Consider the set of primes $v \in \Omega \setminus \Sigma$ such that $\pi^{-1}(P)$ has no smooth k_v-point. By the Lang–Weil theorem combined with Hensel's lemma this set is finite. Let Σ_P be its subset obtained by deleting the primes v such that the image of b is a square in $\mathbf{F}_v$.

We claim that there exists a smooth point $Q \in T_P \cap W'(k)$ such that $H = \pi^{-1}(PQ)$ has points everywhere locally. The equations of $H \subset \mathbf{P}^5_k$ can be written as

$$U_5^2 - bV_5^2 = \varphi_5(Z_1, Z_2), \quad U_6^2 - bV_6^2 = \varphi_6(Z_1, Z_2), \qquad (7.14)$$

where φ_5 and φ_6 are quadratic forms with coefficients in $k(e_5)$ and $k(e_6)$, respectively, which are conjugate over k. The coordinates Z_1 and Z_2 on the line (PQ) are chosen so that P has coordinates $(1 : 0)$, and Q coordinates $(0 : 1)$. We also claim that Q could be chosen so that $\varphi_5(x, 1)\varphi_6(x, 1) \in k[x]$ is an irreducible polynomial of degree 4. (Indeed, φ_5 is irreducible over

$k(e_5)$ for $Q = Q_P$ (it is then easily seen to be proportional to $Z_1^2 - bZ_2^2$), and the same argument works for φ_6. The closed subsets of $T_P \cap X$ defined by $u - e_5 v = 0$ and $u - e_6 v = 0$ are distinct, hence for a point Q outside a certain closed subset, the conjugate polynomials $\varphi_5(x, 1)$ and $\varphi_6(x, 1)$ have no common root. Thus the generic fibre of the covering of $T_P \cap X$ given by $\varphi_5(x, 1)\varphi_6(x, 1)$ is irreducible, and we may hope to apply the Hilbert irreducibility theorem.) Note that if b is a square in $\mathbf{F}_v$, then the reduction of H obviously has smooth $\mathbf{F}_v$-points, hence, by Hensel's lemma, H is locally soluble at such primes v. It remains to ensure that H is locally soluble at primes in Σ_P, because H satisfies the Hasse principle and weak approximation by Corollary 7.3.2. Then Y will have the same properties.

If $\tilde{P}$ belongs to $\Pi_+ \cup \Pi_-$ (case *(a)* above), then since these two projective spaces are disjoint and conjugate over $k(\sqrt{b})$, this would imply that $\tilde{b}$ is a square in $\mathbf{F}_v$. This is not possible for $v \in \Sigma_P$. If $\tilde{P}$ is such that exactly one monomial $x_i y_j y_k \not\equiv 0 \bmod m_v$ (case *(b)*), then we arrive at the same conclusion. There remains the case when m_v divides U_i, V_i, U_j, V_j (case *(c)*), but then the square of m_v divides (u, v), and we have exluded this by our choice of P based on Lemma 7.4.2. Thus for all $v \in \Sigma_P$ the reduction $\tilde{P}$ is a point of $\tilde{W}$. In particular, $x_i y_i \equiv 0 \bmod m_v$ for at most one i. Then, by our choice of Σ, for every $v \in \Sigma_P$ there is a smooth point $Q^v \in T_P \cap X\mathbf{F}_v$ such that $\tilde{F}(Q^v) \neq 0$. Then the fibre $\pi^{-1}(Q^v)$ is a smooth projective rational surface over a finite field, and, as any such surface, contains a rational point. By Lemma 7.4.5, $T_P \cap X$ satisfies weak approximation, hence we can find a k-point Q whose reduction modulo m_v is Q^v for $v \in \Sigma_P$. By Ekedahl's theorem ([E], Thm. 1.2) Q can at the same time be chosen to satisfy the condition of the Hilbert irreducibility theorem, in our case we ensure that the restriction of the polynomial F to the line PQ is irreducible. (This is legitimate since the generic fibre of this covering is irreducible, as we have seen in the previous paragraph.) A smooth $\mathbf{F}_v$-point in the fibre $\pi^{-1}(Q^v)$ lifts to a k_v-point in the fibre $\pi^{-1}(Q)$ by Hensel's lemma. Therefore $\pi^{-1}(Q)(k_v) \neq \emptyset$ for all $v \in \Sigma_P$, which implies that $H = \pi^{-1}(PQ)$ has points everywhere locally.

The proof of the theorem is now complete. QED

If $g(x)$ is not irreducible, then Lemma 7.4.5 is not true any more. The author could not follow the unirationality argument used at the end of [SD99] to overcome this difficulty. There are other ways to do this; however, the exposition of the relevant technique would lead us too far.

Comments

The first three sections are based on [CSS87]. This voluminous paper remains an indispensable reference for the Hasse principle and weak approximation on varieties given by two quadratic equations. Some of the proofs can now be streamlined, some computational 'miracles' explained and some results generalized; see [C90], [CS92], [H94], [H95], [Sl88], [Sl89], [SS], [S90a], [S90b], [S96]. A state of the art account of the arithmetic theory of cubics and intersections of two quadrics requires a book of its own. Let us just quote another unsurpassed result of [CSS87]: smooth proper models of the intersections of two quadrics in $\mathbf{P}_k^8$ satisfy the Hasse principle and weak approximation. A proof of this theorem based on the circle method is not known. At the end of [CSS87] the reader will find various counter-examples to the Hasse principle on intersections of two quadrics.

Theorem 7.3.1 was proved in [CSS87], but when φ_1 and φ_2 are irreducible, the Hasse principle part was already proved in [CCS]. In this case the statement is simply that the *smooth* Hasse principle holds. The proof of Theorem 7.3.1 given here is new. See [CSS97], Example 9.4.1, for a treatment of equations more general than (7.6).

Our proof of Swinnerton-Dyer's theorem in the last section is a modification of his original proof.

An important ingredient of the proofs of Theorems 7.2.1 and 7.4.1 is the so called *fibration method*. For a variety equipped with a morphism to the affine space it sometimes allows one to prove the Hasse principle and weak approximation for the total space once these properties hold for 'sufficiently many' k-fibres (see [S90b]). A more sophisticated method due to D. Harari sometimes allows one to show that the Manin obstruction to the Hasse principle and weak approximation is the only one for the total space once the same condition is true for the k-fibres ([H95], [H97]). The most restrictive condition for the fibration method is that all the fibres must be geometrically integral (though it can be somewhat relaxed). See [S96], [CS00] for the study of the Manin obstruction on a variety fibred over $\mathbf{P}_k^1$ with at most two or three closed geometric fibres which are not integral. These papers use the fibration method and the descent with torsors associated with the morphism to $\mathbf{P}_k^1$ as analysed in Section 4.4. See [CSS98a] for another approach.

8

Non-abelian descent on bielliptic surfaces

The first aim of this chapter is to construct a smooth proper surface X over $k = \mathbf{Q}$ of Kodaira dimension 0 which is a counter-example to the Hasse principle but for which the Manin obstruction is not sufficient to explain the absence of $\mathbf{Q}$-rational points. Then we interpret this example in terms of torsors under finite non-abelian groups, thereby showing that the descent obstruction to the Hasse principle related to torsors under non-abelian groups can be stronger than the Manin obstruction.

We obtain our surface X as the quotient of a product of two curves of genus 1 Y by a fixed point free involution. Numerically such a counter-example can be given by the affine equations

$$(x^2 + 1)y^2 = (x^2 + 2)z^2 = 3(t^4 - 54t^2 - 117t - 243).$$

Geometrically this is a bielliptic surface of type (1) in the list of Bagnera and de Franchis ([Beauville], VI.20). To prove that there is no $\mathbf{Q}$-point we first consider the adelic points of the twisted forms of the torsor $f : Y \to X$ under $\mathbf{Z}/2$, and then use the Manin obstruction for these twisted torsors. The behaviour of the Brauer group here is quite different from the case of an isogeny of abelian varieties: $\mathrm{Br}(Y)$ is substantially bigger than $f^*\mathrm{Br}(X)$. Consequently, the Manin obstruction considered on all the twisted forms of $Y \to X$ gives more restrictions on the existence of $\mathbf{Q}$-points than the Manin obstruction on X.

After some geometric preparations we construct our example at the end of Section 8.1 (Theorem 8.1.5). The construction of the example relies on the existence of elliptic curves over $\mathbf{Q}$ with no rational 2-torsion and an element of exact order 4 in the Tate–Shafarevich group. In the appendix to Section 8.1 (based on computations of S. Siksek) we write down explicitly an everywhere locally soluble 4-covering of the curve $y^2 = x^3 - 1221$, and show that it does represent an element of exact order 4. We observe that our counter-example to the Hasse principle can be explained by the descent obstruction related to an X-torsor under a finite nilpotent group. This interpretation is given in Section 8.2.

8.1 Beyond the Manin obstruction

Let k be a field of characteristic 0. We construct a family of smooth proper surfaces over k. A counter-example to the Hasse principle not accounted for by the Manin obstruction which we intend to build later on in this section will belong to this family.

The construction depends on two pieces of data:

(1) an elliptic curve E over k, a 2-covering $\psi : C \to E$ which lifts to a 4-covering $\psi' : C' \to E$, and such that C has a 0-cycle of degree 2 over k;

(2) an unramified double covering $\phi : D \to D'$ of curves of genus 1.

Let $\xi : C' \to C$ be the morphism such that $\psi' = \psi \circ \xi$. This is a Galois covering with Galois group $E[2]$. In particular, the inverse image on the Jacobians, $\xi^* : E \to E$, is multiplication by 2. A divisor of degree 2 on C defines a morphism $h : C \to \mathbf{P}^1_k$ which makes C a double covering of $\mathbf{P}^1_k$. Let $\sigma : C \to C$ be the corresponding hyperelliptic involution. In particular, σ_* acts on the Jacobian E of C as multiplication by -1.

Let $\rho : D \to D$ be the fixed point free involution interchanging the sheets of the covering $\phi : D \to D'$. Let J (resp. J') be the Jacobian of D (resp. D'). The action of ρ_* on the Jacobian of D is trivial. It is clear that $\phi^* : J' \to J$ is an isogeny of degree 2.

Let X be the quotient of $Y = C \times D$ by the fixed point free involution (σ, ρ), $f : Y \to X$. This is a smooth, proper, geometrically integral surface classically known as hyperelliptic (or bielliptic) (see [Shafarevich], Ch. VII.8, [Beauville], Ch. VI). Its geometric invariants are $\kappa = 0$, $p_g = 0$, $q = 1$, $(K_X^2) = 0$, $b_1 = b_2 = 2$. (Recall that surfaces with such invariants together with $K3$, Enriques and abelian surfaces exhaust all minimal surfaces of Kodaira dimension 0 over an algebraically closed field.)

Finally, let $Y' = C' \times D$, and let $\pi : Y' \to D$ be the second projection. Let $f' : Y' \to X$ be the composition of the unramified coverings $(\xi, Id) : Y' \to Y$ and $f : Y \to X$.

Theorem 8.1.1 *Assume that k is such that $H^3(k, \mathbf{G}_m) = 0$. If E has no k-point of order exactly 2, then the inverse image $f'^*(\mathrm{Br}(X)) \subset \mathrm{Br}(Y')$ is contained in $\pi^*(\mathrm{Br}(D))$.*

Proof. We have $H^0(\overline{X}, \mathbf{G}_m) = \overline{k}^*$ because X is proper, geometrically connected and geometrically reduced. Then the exact sequence (2.23) gives

$$0 \to \mathrm{Br}_0(X) \to \mathrm{Br}_1(X) \to H^1(k, Pic(\overline{X})) \to 0,$$

where we used the assumption $H^3(k, \mathbf{G}_m) = 0$.

For smooth projective surfaces the structure of $\mathrm{Br}(\overline{X})$ was determined by Grothendieck ([Grothendieck 68], II, Cor. 3.4, III, (8.12)): the divisible subgroup $\mathrm{Br}(\overline{X})_{div}$ is isomorphic to $(\mathbf{Q}/\mathbf{Z})^{b_2 - \rho}$ as an abelian group, and the quotient $\mathrm{Br}(\overline{X})/\mathrm{Br}(\overline{X})_{div}$ is isomorphic to $Hom(NS(\overline{X})_{tors}, \mathbf{Q}/\mathbf{Z})$ as a Γ_k-module. Here $b_2 = rk(H^2(\overline{X}, \mathbf{Q}_l))$, $\rho = cork(Pic(\overline{X}) \otimes \mathbf{Q}/\mathbf{Z})$, and $NS(\overline{X})$ is the Néron–Severi group of $\overline{X}$, finitely generated and isomorphic to the quotient of $Pic(\overline{X})$ by its divisible subgroup. Since in our case the geometric genus of X is 0, we have $b_2 = \rho$, hence $\mathrm{Br}(\overline{X})$ is dual to the torsion subgroup of $NS(\overline{X})$.

We now analyse the structure of the Γ_k-module $Pic(\overline{X})$ and of the map $f^* : Pic(\overline{X}) \to Pic(\overline{Y})$. Consider the following commutative diagram:

$$
\begin{array}{ccccc}
C & \xleftarrow{\ \pi_1\ } & Y & \xrightarrow{\ \pi_2\ } & D \\
{\scriptstyle h}\downarrow & & {\scriptstyle f}\downarrow & & {\scriptstyle \phi}\downarrow \\
\mathbf{P}^1_k & \longleftarrow & X & \longrightarrow & D'
\end{array}
$$

The Γ_k-module $J'(\overline{k})$ is identified with $Pic^0(\overline{X})$. Let $F \simeq \mathbf{Z}/2$ be the cyclic group of automorphisms of Y generated by (σ, ρ). Since $f^* : Pic(\overline{X}) \to Pic(\overline{Y})$ factors through the inclusion $Pic(\overline{Y})^F \hookrightarrow Pic(\overline{Y})$ we get a commutative diagram of Γ_k-modules with exact rows:

$$
\begin{array}{ccccccccc}
0 & \to & J'(\overline{k}) & \to & Pic(\overline{X}) & \to & NS(\overline{X}) & \to & 0 \\
 & & {\scriptstyle (0,\phi^*)}\downarrow & & {\scriptstyle f^*}\downarrow & & {\scriptstyle f^*}\downarrow & & \\
0 & \to & E[2](\overline{k}) \times J(\overline{k}) & \to & Pic(\overline{Y})^F & \to & NS(\overline{Y})^F & & \qquad (8.1) \\
 & & \downarrow & & \downarrow & & \downarrow & & \\
0 & \to & E(\overline{k}) \times J(\overline{k}) & \to & Pic(\overline{Y}) & \to & NS(\overline{Y}) & \to & 0
\end{array}
$$

We used the fact that (σ_*, ρ_*) acts on the product of two Jacobians $E(\overline{k}) \times J(\overline{k})$ as multiplication by $(-1, 1)$.

Lemma 8.1.2 *There is an isomorphism of Γ_k-modules*

$$NS(\overline{X})_{tors} = E[2](\overline{k}).$$

Proof. Let us denote the kernel of $f^* : NS(\overline{X}) \to NS(\overline{Y})$ by K. Since $NS(\overline{Y})$ is the Néron–Severi group of an abelian variety it is torsion free. Therefore $NS(\overline{X})_{tors} = K_{tors}$. Thus it will be enough to show that the

Γ_k-modules K and $E[2]$ are isomorphic. The Hochschild–Serre spectral sequence (Section 2.2, Example 1)

$$H^p(F, H^q(\overline{Y}, \mathbf{G}_m)) \Rightarrow H^{p+q}(\overline{X}, \mathbf{G}_m)$$

yields an exact sequence

$$0 \to H^1(F, \overline{k}^*) \to Pic(\overline{X}) \xrightarrow{f^*} Pic(\overline{Y})^F \to H^2(F, \overline{k}^*).$$

We have $H^1(F, \overline{k}^*) = Hom(F, \overline{k}^*) = \mathbf{Z}/2$, and $H^2(F, \overline{k}^*) = 0$ by periodicity of the cohomology of cyclic groups. On the other hand, $Ker(\phi^*) = \mathbf{Z}/2$. The snake lemma applied to the upper part of diagram (8.1) now gives rise to the exact sequence of Γ_k-modules

$$0 \to \mathbf{Z}/2 \to \mathbf{Z}/2 \to K \to E[2] \to 0.$$

This proves Lemma 8.1.2. QED

Corollary 8.1.3 *We have* $\mathrm{Br}(\overline{X})^{\Gamma_k} = 0$, *thus* $\mathrm{Br}(X) = \mathrm{Br}_1(X)$.

Proof. The Weil pairing $E[2] \times E[2] \to \mu_2 = \mathbf{Z}/2$ is Γ_k-invariant and non-degenerate, hence the Γ_k-module $E[2]$ is self-dual. By the second assumption of the theorem we have $E[2]^{\Gamma_k} = 0$. Now Lemma 8.1.2 implies that $\mathrm{Br}(\overline{X}) = Hom(NS(\overline{X})_{tors}, \mathbf{Q}/\mathbf{Z}) = Hom(E[2](\overline{k}), \mathbf{Q}/\mathbf{Z}) = E[2](\overline{k})$ has no non-zero Γ_k-invariant element. QED

Lemma 8.1.4 *The group*

$$f^*(\mathrm{Br}(X))/\mathrm{Br}_0(Y) \subset \mathrm{Br}_1(Y)/\mathrm{Br}_0(Y) = H^1(k, Pic(\overline{Y}))$$

is contained in the image of $H^1(k, E)[2] \times H^1(k, J)$ *under the map induced by the natural inclusion* $E(\overline{k}) \times J(\overline{k}) = Pic^0(\overline{Y}) \hookrightarrow Pic(\overline{Y})$.

Proof. By functoriality of the spectral sequence (2.23) we have to consider the subgroup of $\mathrm{Br}_1(Y)/\mathrm{Br}_0(Y) = H^1(k, Pic(\overline{Y}))$ isomorphic to the image of the map

$$f^* : H^1(k, Pic(\overline{X})) \to H^1(k, Pic(\overline{Y})).$$

Observe that $NS(\overline{Y})^F \subset NS(\overline{Y})$ is torsion free because $\overline{Y}$ is an abelian surface. Since f is a finite covering of degree 2 we have $f_* f^*(R) = 2R$ for any divisor $R \in Div(\overline{X})$, and $S + (\sigma, \rho)S = f^* f_*(S)$ for any divisor $S \in Div(\overline{Y})$. Therefore the Γ_k-modules $NS(\overline{X}) \otimes \mathbf{Q}$ and $NS(\overline{Y})^F \otimes \mathbf{Q}$ are isomorphic. Since $\overline{X}$ has second Betti number $b_2 = 2$, we have $dim(NS(\overline{Y})^F \otimes \mathbf{Q}) = 2$. The classes of the fibres of the canonical projections $\pi_1 : \overline{Y} \to \overline{C}$ and

$\pi_2 : \overline{Y} \to \overline{D}$ give two linearly independent Γ_k-invariant elements of this vector space, implying that it carries trivial Γ_k-action. Thus $NS(\overline{Y})^F = \mathbf{Z} \oplus \mathbf{Z}$ as a Γ_k-module. Consider diagram (8.1) again. Note that on replacing $NS(\overline{Y})^F$ by the image of $Pic(\overline{Y})^F \to NS(\overline{Y})^F$ we obtain from (8.1) a commutative diagram whose middle row is right exact. This image is a submodule of $NS(\overline{Y})^F = \mathbf{Z} \oplus \mathbf{Z}$, and hence is a free abelian group with trivial Γ_k-action. Since $H^1(k, \mathbf{Z}) = 0$ the modified diagram (8.1) gives rise to the following commutative diagram with exact middle row:

$$
\begin{array}{ccccc}
 & & H^1(k, Pic(\overline{X})) & & \\
 & & \downarrow & & \\
H^1(k, E[2]) \times H^1(k, J) & \to & H^1(k, Pic(\overline{Y})^F) & \to & 0 \\
\downarrow & & \downarrow & & \\
H^1(k, E) \times H^1(k, J) & \to & H^1(k, Pic(\overline{Y})) & &
\end{array}
$$

The statement of Lemma 8.1.4 follows from the commutativity of this diagram. QED

Now we can finish the proof of the theorem.

We claim that the group

$$f'^*(\mathrm{Br}(X))/\mathrm{Br}_0(Y') \subset \mathrm{Br}_1(Y')/\mathrm{Br}_0(Y) = H^1(k, Pic(\overline{Y'}))$$

is contained in the image of $H^1(k, J)$ under the map induced by the natural inclusion $E(\overline{k}) \times J(\overline{k}) = Pic^0(\overline{Y'}) \hookrightarrow Pic(\overline{Y'})$. After Lemma 8.1.4 and by functoriality of the spectral sequence (2.23) in order to prove this we only have to remark that the inverse image map $Pic^0(\overline{Y}) \to Pic^0(\overline{Y'})$ is multiplication by $(2, 1)$ on $E(\overline{k}) \times J(\overline{k})$. The theorem is proved. QED

Now we construct our counter-example.

Let E be an elliptic curve over $\mathbf{Q}$ whose Tate–Shafarevich group $\mathrm{III}(E)$ contains an element of order 4, and such that E contains no non-trivial point of order 2 defined over $\mathbf{Q}$. For example, take E to be the curve

$$y^2 = x^3 - 1221.$$

Then $E(\mathbf{Q}) = 0$ and $\mathrm{III}(E) = \mathbf{Z}/4 \times \mathbf{Z}/4$ (the last property is conditional on the Birch–Swinnerton-Dyer conjecture, see [GPZ], Tables 3 and 4). We shall only use the following (unconditional) result proved in the appendix to this section. The curve C given by

$$u^2 = g(t), \quad \text{where} \quad g(t) = 3(t^4 - 54t^2 - 117t - 243),$$

can be equipped with the structure of a 2-covering of E, which lifts to a 4-covering $C' \to E$ such that $[C'] \in \mathrm{III}(E)$, and on the other hand $C(\mathbf{Q}) = \emptyset$.

Choose two monic quadratic polynomials $p(x)$ and $q(x)$ with integer coefficients such that their resultant is ± 1, and such that both $p(x)$ and $q(x)$ take only positive values on $\mathbf{Q}$. (For instance, $p(x) = x^2+1$, $q(x) = x^2+2$.) Let $D \subset \mathbf{P}^3_{\mathbf{Q}}$ be the smooth proper curve of genus 1 given by its affine equations

$$y^2 = p(x), \ z^2 = q(x).$$

This curve D has obvious rational points at infinity. Let $\rho : D \to D$ be the involution sending (x, y, z) to $(x, -y, -z)$. One sees easily that ρ has no fixed point. Then $D' = D/\rho$ is an elliptic curve given by

$$w^2 = p(x)q(x).$$

Let $\phi : D \to D'$ be the natural surjective map.

As before we define X as the quotient of $Y = C \times D$ by the fixed point free involution (σ, ρ), $f : Y \to X$. An affine model of X can be given by equations

$$y^2 = g(t)p(x), \ z^2 = g(t)q(x).$$

We keep the same notation as before: $Y' = C' \times D$, $\pi : Y' \to D$ is the second projection, and $f' : Y' \to X$ is the composition $f \circ (\xi, Id)$.

The covering $f : Y \to X$ is clearly a torsor under $\mathbf{Z}/2$. We have $H^1(\mathbf{Q}, \mathbf{Z}/2) = \mathbf{Q}^*/\mathbf{Q}^{*2}$. According to the general theory, $X(\mathbf{Q})$ is the union of images $f^a(Y^a(\mathbf{Q}))$, $a \in \mathbf{Q}^*$, where $f^a : Y^a \to X$ is the twist of f by the cocycle given by a. Since f is given by extracting the square root of $g(t)$, the twisted variety Y^a is the product of the twisted curve C^a, given by $y^2 = ag(t)$, and the twisted curve D^a, given by $y^2 = ap(x)$, $z^2 = aq(x)$.

Theorem 8.1.5 *(a) With notation as above we have $X(\mathbf{Q}) = \emptyset$.*

(b) For any $R \in D(\mathbf{Q})$, and any $\{P_v\} \in C'(\mathbb{A}_{\mathbf{Q}})$, the map f' sends $\{(P_v, R)\} \in Y'(\mathbb{A}_{\mathbf{Q}})$ to $X(\mathbb{A}_{\mathbf{Q}})^{\mathrm{Br}}$; in particular, we have $X(\mathbb{A}_{\mathbf{Q}})^{\mathrm{Br}} \neq \emptyset$.

Proof. (a) An easy valuation argument shows that for any prime p and any $\mathbf{Q}_p$-rational point of X for which $yz \neq 0$ the p-adic valuation $val_p(g(t))$ is even. (Indeed, if $val_p(x) \geq 0$, then since the resultant of $p(x)$ and $q(x)$ is ± 1, either $val_p(p(x))$ or $val_p(q(x))$ equals 0. The equations then show that $val_p(g(t))$ is even. If $val_p(x) < 0$, then since $p(x)$ and $q(x)$ are quadratic with integer coefficients, both $val_p(p(x))$ and $val_p(q(x))$ are even, hence $val_p(g(t))$ is even.) Thus $g(t) = \pm 1$ modulo squares in $\mathbf{Q}^*$. This means that if $Y^a(\mathbf{Q}) \neq \emptyset$, then $a = \pm 1$. We get a decomposition

$$X(\mathbf{Q}) = f(Y(\mathbf{Q})) \cup f^-(Y^-(\mathbf{Q})),$$

where $f^- : Y^- = C^- \times D^- \to X$ is a 'twisted form' of f, and C^- (resp. D^-) is obtained by inverting the sign of $g(t)$ (resp. of $p(x)$ and $q(x)$). Note that $Y = C \times D$ clearly has no rational point since $C(\mathbf{Q})$ is empty. The curve D^- given by the equations $y^2 + p(x) = z^2 + q(x) = 0$ has no real point by the positivity condition on $p(x)$ and $q(x)$, hence Y^- has no real point. This completes the proof of *(a)*.

(b) By the projection formula it is enough to show that $\{(P_v, R)\}$ is Brauer–Manin orthogonal to $f'^*(\mathrm{Br}(X))$, and this follows from Theorem 8.1.1 by the global reciprocity law. QED

Remark. J.-L. Colliot-Thélène conjectured that the Manin obstruction to the Hasse principle for 0-cycles of degree 1 is the only obstruction for all varieties over a number field k ([C95], Conj. 1.5 (a); this statement was also formulated and discussed by S. Saito in [Saito], Sect. 8). If this conjecture is true, then for some odd degree extension $K/\mathbf{Q}$ there would exist a K-point on X. I do not know how to verify this.

Appendix. An example of 4-torsion in $\text{III}(E)$. Let us apply Proposition 3.3.6 to the curve

$$C: \ y^2 = g(x), \quad g(x) = 3(x^4 - 54x^2 - 117x - 243)$$

defined over $\mathbf{Q}$. Using statement *(a)* of that proposition one computes that this is a 2-covering of

$$E: \ y^2 = x^3 - 1221.$$

The curve C is in fact everywhere locally soluble, and was initially found by S. Siksek using Cremona's program `mwrank` [Cremona]. The computations in this appendix are due to him.

Note in passing that by [BSD], Lemmas 1 and 2, [Cassels 62], IV, Thm. 1.3, any 2-covering which is everywhere locally soluble can be given by a double cover of $\mathbf{P}_{\mathbf{Q}}^1$.

It is computed in [GPZ] that E has analytic rank 0. Then the rank of $E(\mathbf{Q})$ is unconditionally 0 by the work of Rubin and Kolyvagin. By a theorem of Rubin ([R], Thm. A, Remark 3, p. 528) the Tate–Shafarevich group $\text{III}(E)$ is finite. The classical computation of torsion of such curves implies that $E(\mathbf{Q}) = \{0\}$. This together with irreducibility of $g(x)$ implies that $C(\mathbf{Q}) = \emptyset$. It is also computed in [GPZ] that $\text{III}(E)$ is predicted to be isomorphic to $\mathbf{Z}/4 \times \mathbf{Z}/4$ by the conjectures of Birch and Swinnerton-Dyer. We shall now exhibit an everywhere locally solvable 4-covering $C' \to E$ which is a lifting of the 2-covering $C \to E$. By *(c)* of Proposition 3.3.6 this gives an element of order exactly 4 in $\text{III}(E)$.

It is observed that the element $\epsilon = -\theta^3/3 - \theta^2 + 29\theta + 27 \in K = \mathbf{Q}(\theta)$, where $g(\theta) = 0$, has norm $243 = 3 \times 9^2$. We construct a 4-covering C' as the intersection of two quadrics as in Proposition 3.3.6 *(b)*. The corresponding quadratic forms can be written down explicitly:

$$\mathbf{x}A\mathbf{x}^t = 0, \quad \mathbf{x}B\mathbf{x}^t = 0$$

where $\mathbf{x} = (x_1, x_2, x_3, x_4)$ and the entries of A and B are respectively

$$\begin{pmatrix} -1 & 11 & -66 & 396 \\ 11 & -66 & 396 & -2520 \\ -66 & 396 & -2520 & 16,335 \\ 396 & -2520 & 16,335 & -105,786 \end{pmatrix} \text{ and } \begin{pmatrix} -1 & -3 & 33 & -198 \\ -3 & 33 & -198 & 1188 \\ 33 & -198 & 1188 & -7560 \\ -198 & 1188 & -7560 & 49,005 \end{pmatrix}.$$

Let us show that $[C'] \in \mathrm{III}(E)$. Criteria for testing intersections of two quadrics in $\mathbf{P}^3$ for everywhere local solubility are given in [MSS]. By Lemma 7 of [MSS] we know that it is soluble over $\mathbf{R}$, and by Thm. 4 of [MSS] that it is necessary to test for solubility only at the finite primes 2, 3, 11, 37 (2 and the divisors of 1221). This result also tells us that the following will lift to p-adic points on C' for $p = 2$, 3, 11, 37, respectively: $(0, 2, 1, 0)$ mod 2^3, $(12, 21, 1, 0)$ mod 3^3, $(0, 1, 0, 0)$ mod 11, $(0, 1, 9, 16)$ mod 37. This completes our proof that C' represents an element of order exactly 4 in the Tate–Shafarevich group of E.

8.2 Interpretation in terms of non-abelian torsors

We keep the notation of the previous section. Let C'' be an 8-covering of E which is a lifting of the 4-covering $C' \to E$. To produce it we have to divide the class $[C'] \in H^1(k, E)$ by 2. This can be done because any element of $\mathrm{III}(E)$ is divisible by a prime number in $H^1(k, E)$ if E is an elliptic curve (see [Milne, ADT], I.6.18). Let the morphisms $\xi' : C'' \to C'$ and $\xi : C' \to C$ be induced by multiplication by 2 in E. Let us consider the tower of finite étale coverings

$$Y'' = C'' \times D \xrightarrow{\xi' \times Id} Y' = C' \times D \xrightarrow{\xi \times Id} Y = C \times D \xrightarrow{f} X.$$

Let $f' = f \circ (\xi \times Id)$, $f'' = f \circ (\xi \times Id) \circ (\xi' \times Id)$. The map f'' makes Y'' an X-torsor with respect to the k-group G defined as the semi-direct product of $E[4]$ by $\mathbf{Z}/2$, where the non-trivial element of $\mathbf{Z}/2$ acts on $E[4]$ as multiplication by -1. Suppose that $[C] \in \mathrm{III}(E)[2]$ is not divisible by 4 in $\mathrm{III}(E)$. (It follows from the Birch and Swinnerton-Dyer conjecture that $\mathrm{III}(E) = (\mathbf{Z}/4)^2$, so this should be the case here.)

Proposition 8.2.1 *Assume that the Birch and Swinnerton-Dyer conjecture is true for the elliptic curve E. Then no twisted form of the torsor f'' : $Y'' \to X$ has points everywhere locally.*

Note that we have shown in Theorem 8.1.5 that $X(\mathbb{A}_k)^{\mathrm{Br}} \neq \emptyset$, thus the proposition means that the obstruction to the Hasse principle associated to a torsor under a non-abelian group can be stronger than the Manin obstruction.

Without assuming the Birch–Swinnerton-Dyer conjecture the same result remains true if one chooses $[C'] \in \mathrm{III}(E)$ such that $[C] = 2^i[C']$ with i maximal with this property, and then defines C'' and Y'' in the same way as above. (The group $\mathrm{III}(E)$ is finite by Rubin's theorem.)

Proof of proposition. Let $\sigma \in Z^1(k, G)$, and let $(Y'')^\sigma \to X$ be the twist of $Y'' \to X$ by σ. This is an X-torsor under the twisted k-group G^σ. This group contains $E[4]^\sigma$ as a normal k-subgroup of index 2. The quotient of $(Y'')^\sigma$ by $E[4]^\sigma$ is an X-torsor under $\mathbf{Z}/2$. This is the same thing as the twist Y^τ of $Y \to X$ by the image $\tau \in Z^1(k, \mathbf{Z}/2)$ of σ. In the proof of Theorem 8.1.5 we have shown that if $Y^\tau(\mathbf{A_Q}) \neq \emptyset$, then $[\tau] = 1$ and Y^τ is isomorphic to Y. Hence the morphism $(Y'')^\sigma \to X$ factors as $(Y'')^\sigma \to Y \to X$. From the exact sequence of pointed sets

$$H^1(k, E[4]) \to H^1(k, G) \to H^1(k, \mathbf{Z}/2)$$

it now follows that there exists $\rho \in Z^1(k, E[4])$ such that $[\sigma]$ is the image of $[\rho]$. Then $(Y'')^\sigma \to Y$ can be considered as the twist of $Y'' = C'' \times D \to Y$ by ρ. The action of $E[4]$ on $Y'' = C'' \times D$ is given by the natural action on C'' and the trivial action on D. Thus $(Y'')^\sigma = (C'')^\rho \times D$. It is clear that $(C'')^\rho$ is an 8-covering of E which is a lifting of the 2-covering $C \to E$. Since $\mathrm{III}(E) = (\mathbf{Z}/4)^2$ by the Birch–Swinnerton-Dyer conjecture, we conclude that $(C'')^\rho$ cannot represent an element of $\mathrm{III}(E)$ for any class $[\rho] \in H^1(k, E[4])$. Thus no twist of $Y'' \to X$ has an adelic point. QED

Counter-examples to weak approximation not explained by the Brauer–Manin obstruction.

Let us quote the following general criterion for a variety over a number field to be a counter-example to weak approximation not explained by the Brauer–Manin obstruction. As in the previous counter-example to the Hasse principle, the tool which allows one to get a better 'bound' for $X(k)$ inside $X(\mathbb{A}_k)$ than just $X(\mathbb{A}_k)^{\mathrm{Br}}$ is the non-abelian descent obstruction.

Theorem 8.2.2 (D. Harari) *Let X be a proper, smooth and geometrically integral variety over a number field k such that $X(k) \neq \emptyset$. Assume that $\mathrm{Br}(\overline{X})^{\Gamma_k}$ is finite (e.g. $H^2(X, \mathcal{O}_X) = 0$) and that the geometric fundamental group $\pi_1(\overline{X})$ is non-abelian. Assume further that $H^1(X, \mathcal{O}_X) = 0$ or that the tangent bundle of X is numerically effective. Then the closure of $X(k)$ in $X(\mathbb{A}_k)$ is strictly smaller than $X(\mathbb{A}_k)^{\mathrm{Br}}$.*

This result is based on the descent obstruction to weak approximation defined in Section 5.3. What happens is that the assumption that $\pi_1(\overline{X})$ is non-abelian gives a torsor $f : Y \to X$ under a finite non-abelian group scheme G. Now using the abelian descent theory of Section 6.1 one shows that infinitely many elements of $X(\mathbb{A}_k)^{\mathrm{Br}}$ do not belong to $X(\mathbb{A}_k)^f$. Thus the fact that G is not abelian implies that the descent obstruction to weak approximation can be finer that the Brauer–Manin obstruction. An interesting feature of the above theorem is that all the hypotheses (except $X(k) \neq \emptyset$) are purely geometric. In particular, a counter-example of this kind will remain one after a finite extension of the ground field.

This result applies to étale quotients of abelian varieties, in particular, to bielliptic surfaces (this is not surprising in view of the previous counter-example to the Hasse principle). It is worth noting that in this case, $X(K)$ is Zariski dense in X for some finite field extension K/k. We refer the reader to the original paper [H] for details.

Comments

The exposition in this chapter follows [S99] and [HS].

Rational points on bielliptic surfaces were first studied in [CSS97]. The bielliptic surface X defined over $\mathbf{Q}$ by the equations

$$y^2 = (4t^4 + t^2 - 4)x, \quad z^2 = (4t^4 + t^2 - 4)(x^2 + 4x - 1)$$

has an interesting property, that it has two real connected components, and the $\mathbf{Q}$-rational points are dense (in the real topology) in one component, and are present but not dense in the other. The proof is based on the fact that a natural X-torsor under $\mathbf{Z}/2$ is the product of two elliptic curves, say C and D. One shows similarly to the proof of Theorem 8.1.5 that

$$X(\mathbf{Q}) = f(C(\mathbf{Q}) \times D(\mathbf{Q})) \cup f^-(C^-(\mathbf{Q}) \times D^-(\mathbf{Q})),$$

$$X(\mathbf{R}) = f(C(\mathbf{R}) \times D(\mathbf{R})) \cup f^-(C^-(\mathbf{R}) \times D^-(\mathbf{R})),$$

where minus denotes the twist by the cocycle corresponding to the class of -1 in $H^1(\mathbf{Q}, \mathbf{Z}/2) = \mathbf{Q}^*/\mathbf{Q}^{*2}$. The curves C and D in this example are such that $C(\mathbf{R})$, $D(\mathbf{R})$, $C^-(\mathbf{R})$ and $D^-(\mathbf{R})$ are connected, thus the second partition describes the two connected components of X. All these curves have infinitely many $\mathbf{Q}$-rational points, except D^- which has $\mathbf{Q}$-points but only finitely many. Thus $\mathbf{Q}$-rational points are dense in one connected component of X, but in the other component their real closure is non-empty and lies on finitely many curves.

The existence of such a surface contradicts a conjecture of Barry Mazur ([M92], [M95]) who suggested that if the set $X(\mathbf{Q})$ is Zariski dense in a smooth integral variety X, then the real closure of $X(\mathbf{Q})$ is the union of (finitely many) connected components of X. Mazur was interested in these questions in connection with the problems of Diophantine decidability [M94] (see also [Sh97]). A natural modification of Mazur's conjectures that takes into account this and other counter-examples can be found at the end of [CSS97].

A first conditional counter-example to the Hasse principle not explained by the Manin obstruction was constructed by P. Sarnak and L. Wang in [SW]. They have produced an infinite family of smooth hypersurfaces of degree 1130 in $\mathbf{P}_{\mathbf{Q}}^4$ with points everywhere locally. If one assumes Lang's conjecture that $X(\mathbf{Q})$ is finite when $X_{\mathbf{C}}$ is hyperbolic, then only finitely many hypersurfaces in this family have a $\mathbf{Q}$-point.

9

Homogeneous spaces and non-abelian cohomology

The aim of this chapter is to prove that the Manin obstruction to the Hasse principle is the only one for homogeneous spaces of simply connected semisimple groups with connected stabilizer (see Theorem 9.5.1 below for a more precise statement). This is an essential part of a more general theorem of Borovoi valid for any connected affine algebraic group. For the deduction of this more general statement we refer the reader to the original paper [Bor 96]. We follow Borovoi's method; however, the use of universal torsors and gerbs leads to a shorter and more conceptual proof.

We recall the definition of liens and gerbs in Section 9.1, and Springer's construction of the second non-abelian cohomology class attached to a homogeneous space in Section 9.2. An accessible exposition of second non-abelian cohomology deserves a book of its own. In this chapter we have to content ourselves with a very sketchy description of basic properties of gerbs and liens relevant to our purposes. In Section 9.3 following Borovoi [Bor 93] we define the abelianization of the second cohomology. The arithmetical applications are contained in the last two sections. The reader can complement the exposition in this chapter by looking in [Giraud], [DM], [FSS], [Bor 93] for more details.

9.1 Liens and non-abelian H^2

In this section k is a field of characteristic 0. Let us denote by $\gamma^* : Spec(\overline{k}) \to Spec(\overline{k})$ the isomorphism induced by $\gamma \in \Gamma_k$. Let $\overline{G}$ be an algebraic $\overline{k}$-group, and $\overline{p} : \overline{G} \to Spec(\overline{k})$ be the structure morphism. We denote by $\mathrm{SAut}(\overline{G})$ the group of *semilinear k-automorphisms* of $\overline{G}$, that is, automorphisms φ of the group scheme $\overline{G}$ such that $\overline{p} \circ \varphi = (\gamma^*)^{-1} \circ \overline{p}$ for some $\gamma \in \Gamma_k$. Such a γ is then unique. The map sending φ to γ is a homomorphism $\mathrm{SAut}(\overline{G}) \to \Gamma_k$; we denote it by q.

168

The group $\mathtt{SAut}(\overline{G})$ is equipped with the weak topology associated to the discrete topology on $\overline{G}(\overline{k})$, that is, the coarsest topology for which the map $\varphi \mapsto \varphi(\overline{g})$ is continuous for any $\overline{g} \in \overline{G}(\overline{k})$. In other words, this is the coarsest group topology on $\mathtt{SAut}(\overline{G})$ for which the stabilizers of geometric points $\overline{g} \in \overline{G}(\overline{k})$ are open.

Lemma 9.1.1 *Let $\overline{G}$ be an algebraic $\overline{k}$-group. The continuous homomorphic sections of $q : \mathtt{SAut}(\overline{G}) \to \Gamma_k$ are in a natural bijective correspondence with k-forms of $\overline{G}$, that is, algebraic k-groups G such that $\overline{G} = G \times_k \overline{k}$.*

Proof. If $\overline{G} = G \times_k \overline{k}$, then the action of Γ_k on $\overline{G}$ via its action on $\overline{k}$ defines a homomorphic section of q. It is continuous because the preimage of the stabilizer of $\overline{g} \in \overline{G}(\overline{k})$ is the open subgroup $\Gamma_{k(\overline{g})} \subset \Gamma_k$, where $k(\overline{g})$ is the field of definition of $\overline{g}$. Conversely, a continuous section of q induces an action of Γ_k on $\overline{G}(\overline{k})$ such that the stabilizer of any point is open. By [BS], 2.12 and [Serre, GA], V.20 we obtain G as the quotient of $\overline{G}$ by this action of Γ_k. QED

Note that q itself need not be continuous.

Let G be an abstract group. An element $g \in G$ induces an inner automorphism of G defined by $(\mathrm{int}\, g)(h) = ghg^{-1}$. We let $Out(G)$ be the quotient of the group $Aut(G)$ of automorphisms of G by the inner automorphisms. An algebraic $\overline{k}$-group $\overline{G}$ gives rise to an exact sequence of groups:

$$1 \to \mathtt{Aut}(\overline{G}) \to \mathtt{SAut}(\overline{G}) \xrightarrow{q} \Gamma_k. \tag{9.1}$$

Here $\mathtt{Aut}(\overline{G})$ is the automorphism group of $\overline{G}$ as an algebraic $\overline{k}$-group. Let $\mathtt{Inn}(\overline{G})$ be the group of inner automorphisms of the algebraic $\overline{k}$-group $\overline{G}$; this is the quotient of G by its centre. We set

$$\mathtt{Out}(\overline{G}) = \mathtt{Aut}(\overline{G})/\mathtt{Inn}(\overline{G}), \quad \mathtt{SOut}(\overline{G}) = \mathtt{SAut}(\overline{G})/\mathtt{Inn}(\overline{G}).$$

The natural action of $\mathtt{SAut}(\overline{G})$ on $\overline{G}(\overline{k})$ induces a canonical map $\mathtt{SOut}(\overline{G}) \to Out(\overline{G}(\overline{k}))$. The sequence (9.1) modulo $\mathtt{Inn}(\overline{G})$ gives rise to

$$1 \to \mathtt{Out}(\overline{G}) \to \mathtt{SOut}(\overline{G}) \to \Gamma_k. \tag{9.2}$$

The following definition is due to Flicker, Scheiderer and Sujatha; cf. [FSS], Def. (1.11).

Definition 9.1.2 *A k-lien on $\overline{G}$ is a splitting $L : \Gamma_k \to \mathtt{SOut}(\overline{G})$ of (9.2), which lifts to a continuous map (not necessarily a homomorphism) $\Gamma_k \to \mathtt{SAut}(\overline{G})$. A lien L is called **representable** if L lifts to a continuous homomorphism $\Gamma_k \to \mathtt{SAut}(\overline{G})$.*

The following properties easily follow from Lemma 9.1.1. A k-form G of $\overline{G}$ defines a representable k-lien which we denote by $lien(G)$. If G and G' are k-forms of $\overline{G}$, then $lien(G) = lien(G')$ if and only if G' is an inner form of G. It is clear that if $\overline{G}$ is abelian, then the k-liens on $\overline{G}$ bijectively correspond to k-forms of $\overline{G}$.

We now define a class of extensions of topological groups in the following unusual sense. Consider an extension of topological groups

$$1 \to \overline{G}(\overline{k}) \to E \xrightarrow{\ q\ } \Gamma_k \to 1 \qquad\qquad (9.3)$$

where $\overline{G}(\overline{k})$ has the discrete and Γ_k its natural profinite topology. We do not however require the arrows to be continuous! Instead we require that (9.3) be *locally split* in the following sense: for a finite field extension K/k the induced map $q_K : E_K \to \Gamma_K$ admits a continuous homomorphic section, where by definition $E_K := E \cap q^{-1}(\Gamma_K)$. Two such extensions are called equivalent if there is a topological isomorphism of their middle terms which induces identity on $\overline{G}(\overline{k})$ and Γ_k.

The condition that an extension (9.3) is locally split is equivalent to the condition that the maps in (9.3) are open onto their images (this last condition is used in [FSS]; the equivalence is proved in [HS], Appendix A).

Let L be a k-lien on $\overline{G}$. An extension (9.3) is called *compatible* with L if the induced homomorphism $\Gamma_k \to Out(\overline{G}(\overline{k}))$ is $L : \Gamma_k \to \mathrm{SOut}(\overline{G})$ followed by the canonical map $\mathrm{SOut}(\overline{G}) \to Out(\overline{G}(\overline{k}))$.

Definition 9.1.3 *Let L be a k-lien on an algebraic $\overline{k}$-group $\overline{G}$.* **The second Galois cohomology set $H^2(k, L)$** *is defined as the set of equivalence classes of locally split extensions of topological groups which are compatible with the lien L. The neutral elements of $H^2(k, L)$ are given by the extensions which admit a continuous homomorphic section.*

Alternatively, $H^2(k, L)$ can be defined in terms of cocycles ([Springer 66], [FSS], (1.17), (1.25), (1.19)).

The set $H^2(k, L)$ may be empty. Note also that in general $H^2(k, L)$ is not functorial in L. For example, if the underlying $\overline{k}$-group of L_1 is trivial, and $H^2(k, L_2) = \emptyset$, then there is no map $H^2(k, L_1) \to H^2(k, L_2)$. However, such a natural map exists if the morphism of liens is surjective, or if the lien L_2 is abelian (see [Giraud], IV.3 or [Bor 93], 1.7).

The subset of neutral elements of $H^2(k, L)$ is denoted by $H^2(k, L)'$. It is non-empty precisely when L is representable. The correspondence between split extensions and k-forms of $\overline{G}$ compatible with L associates to a k-form G the semi-direct product $G(\overline{k}) \rtimes \Gamma_k$ with respect to the natural action of Γ_k on $G(\overline{k})$. Note in passing that the quotient of $\overline{G}$ by the image of a section of $G(\overline{k}) \rtimes \Gamma_k \to \Gamma_k$ is a k-torsor under G.

If G is abelian, then it is easy to check that $H^2(k, lien(G))$ is just the usual Galois cohomology group $H^2(k, G)$, and its unique neutral element is 0 (see also [Giraud], IV.3.4). If G is an algebraic k-group we shall write $H^2(k, G)$ for $H^2(k, lien(G))$. A k-lien is called affine (resp. connected, reductive, semisimple, etc.) when so is its underlying algebraic $\overline{k}$-group.

The second Galois cohomology set is a reasonable number-theoretic object. For example, when k is a non-archimedean local field or a totally imaginary number field, and L is semisimple and simply connected, Douai [D] proved that all the elements of $H^2(k, L)$ are neutral. This can be compared with the classical result that under the same assumptions $H^1(k, G) = 1$, where G is a semisimple and simply connected k-group.

Another useful interpretation of elements of $H^2(k, L)$ is provided by k-*gerbs*. Such a gerb is a category $\mathcal{G}$ fibred over the category of extensions $k \subset K \subset \overline{k}$ such that the fibre $\mathcal{G}(K)$ is a groupoid, and such that $\mathcal{G}(K) \neq \emptyset$ for some finite extension K/k. Moreover, $\mathcal{G}$ must be a stack, that is, must satisfy certain gluing properties. Elements of $\mathcal{G}(K)$ are called sections over K, and a section over k is called a global section. A k-gerb is called *neutral* if it has a global section. An example of a neutral k-gerb is the gerb of torsors $\mathrm{TORS}(G)$ of an algebraic k-group G. By definition, the fibre $\mathrm{TORS}(G)(K)$ is the groupoid of K-torsors under G_K.

Let L be a k-lien on $\overline{G}$, and E be an extension (9.3) compatible with L. We associate to E the gerb $\mathcal{G}_E$ by defining $\mathcal{G}_E(K)$ as the set of continuous sections of $q_K : E_K \to \Gamma_K$. By [Giraud], VIII.6.2.5 and VIII.7.2.5 the cohomology class of E is the cohomology class of the gerb of sections $\mathcal{G}_E$ in the sense of [Giraud]. It is clear from the previous discussion that if G is a k-form of $\overline{G}$, and $E = G(\overline{k}) \rtimes \Gamma_k$ is the natural semi-direct product of $G(\overline{k})$ and Γ_k, then $\mathcal{G}_E = \mathrm{TORS}(G)$. Thus to split extensions (which give rise to neutral cohomology classes) correspond neutral gerbs.

Let us now list the properties which we shall use later in this chapter:

(1.1) Whenever a lien L is abelian, the underlying algebraic $\overline{k}$-group $\overline{G}$ has a uniquely defined k-form G compatible with L, and $L = lien(G)$. In particular, the centre of L (that is, the restriction of L to the centre of $\overline{G}$) is $lien(Z)$ where Z is an abelian algebraic k-group such that $\overline{Z}$ is the centre of $\overline{G}$.

(1.2) If $H^2(k, L)$ is not empty, it is a principal homogeneous space of the cohomology group $H^2(k, Z)$, where $\overline{Z}$ is the centre of L ([Springer 66],

1.17). If $\alpha, \alpha' \in H^2(k, L)$, then we shall denote by $\alpha - \alpha'$ the element of $H^2(k, Z)$ which sends α' to α.

(1.3) Suppose that L is representable, $L = lien(G)$. Then $H^2(k, L)$ contains the class $[\mathrm{TORS}(G)]$. This distinguished class defines a natural identification of $H^2(k, L)$ with $H^2(k, Z)$ which identifies $[\mathrm{TORS}(G)]$ with 0. Then the set $H^2(k, L)'$ identifies with the image of the differential $\delta :$ $H^1(k, G/Z) \to H^2(k, Z)$ associated to the central extension

$$1 \to Z \to G \to G/Z \to 1.$$

Explicitly, if ϕ is a 1-cocycle of the Galois group Γ_k with coefficients in G/Z, then $\delta(\phi)$ corresponds to the class $[\mathrm{TORS}(_\phi G)] \in H^2(k, L)$, where $_\phi G$ is the inner form of G obtained by twisting G by ϕ.

(1.4) If L is connected and reductive, then it is representable. (This is done by using Chevalley's construction; see [Bor 93], Prop. 3.1.)

9.2 The Springer class of a homogeneous space

Let X be a k-variety which is a right homogeneous space of an algebraic k-group G. In general, $X(k)$ could be empty and the stabilizers of $\overline{k}$-points of X are $\overline{k}$-subgroups of $\overline{G}$ which do not necessarily come from k-subgroups of G. However, X defines the structure of a lien L_X on the stabilizer $\overline{H}$ of some point $\overline{x}_0 \in X(\overline{k})$. This lien does not depend on the choice of $\overline{x}_0$. This allows one to define the Galois cohomology set $H^2(k, L_X)$, and the class of X in it, $\eta_X \in H^2(k, L_X)$. It was introduced in [Springer 66].

The class η_X is defined by the extension (cf. [FSS], (5.1))

$$1 \to \overline{H}(\overline{k}) \to E_X \to \Gamma_k \to 1 \tag{9.4}$$

where E_X is the subgroup of $G(\overline{k}) \rtimes \Gamma_k$ consisting of the products $\overline{g}\gamma$ such that $\overline{g}\overline{x}_0 = \gamma(\overline{x}_0)$. Here are some of the properties of this construction:

(2.1) If $X(k) \neq \emptyset$, then η_X is neutral (in this case $E_X = H(\overline{k}) \rtimes \Gamma_k$ where H is the stabilizer of a k-point in X). This implies, in particular, that (9.4) is locally split.

(2.2) The class η_X is neutral if and only if there are a right k-torsor Y under G and a morphism $Y \to X$ commuting with the action of G. (To a section s of $E_X \to \Gamma_k$ one associates the quotient Y of $\overline{G}$ by the image $s(\Gamma_k)$.)

(2.3) In the assumptions of (2.1) and (2.2) let $H = Aut_G(Y/X)$ be the algebraic k-group of automorphisms of Y commuting with the action of G and compatible with the projection to X. In particular, we have

$H(\overline{k}) = Aut_{G(\overline{k})}(\overline{Y}/\overline{X})$. Then X is the quotient of Y by the left action of H, and $L_X = lien(H)$. This makes Y a left X-torsor under H. Such a pair (H, Y) is then unique up to twisting both H and Y by a k-torsor under H.

One can associate to X a k-gerb $\mathcal{G}_X$ whose cohomology class in the sense of [Giraud] is $\eta_X \in H^2(k, L_X)$. If $k \subset K$ is a field extension, then the set of sections $\mathcal{G}_X(K)$ consists of the K-torsors Y_K under G_K equipped with a morphism $Y_K \to X_K$ commuting with the right action of G_K.

We shall only consider the case when L_X is representable. By property (1.4) this is always satisfied when the stabilizers of $\overline{k}$-points of X are connected and reductive (or abelian).

In the most simple case when $H^1(k, G)$ is trivial (e.g. $G = GL_n$, $G = SL_n$, or G is a semisimple simply connected group over a local field or over a totally imaginary number field), we have $X(k) \neq \emptyset$ if and only if the class η_X is neutral in $H^2(k, L_X)$.

9.3 Abelianization of non-abelian H^2

Let L be a connected reductive k-lien. By property (1.4) it can be represented by an algebraic k-group H. Let H^{ss} be the derived group of H (it is semisimple), $H^{tor} = H/H^{ss}$ (it is a torus), H^{sc} be the universal covering of H^{ss} (it is a simply connected semisimple group). Let Z be the centre of H, and Z^{sc} be the centre of H^{sc}.

Following Borovoi we define the group $H^2_{ab}(k, L)$ as the second hypercohomology group of the natural complex of abelian k-groups $(Z^{sc} \to Z)[1]$:

$$H^2_{ab}(k, L) = \mathbb{H}^2(k, (Z^{sc} \to Z)[1]).$$

This gives rise to the following exact sequence of abelian groups:

$$H^2(k, Z^{sc}) \to H^2(k, Z) \to H^2_{ab}(k, L) \to H^3(k, Z^{sc}) \qquad (9.5)$$

Lemma 9.3.1 *Let α and α' be two neutral elements of $H^2(k, L)$; then $\alpha - \alpha'$ considered as an element of $H^2(k, Z)$ (property (1.2)) has zero image in $H^2_{ab}(k, L)$.*

Proof. Let G be a k-form of the underlying $\overline{k}$-group of L corresponding to α'. By property (1.3) the element $\alpha - \alpha' \in H^2(k, Z)$ belongs to the image of the map $\delta : H^1(k, H/Z) \to H^2(k, Z)$. The adjoint group of the connected reductive group H, $H^{ad} = H/Z$, coincides with the adjoint group of the semisimple simply connected group H^{sc}. We thus obtain a central extension

$$1 \to Z^{sc} \to H^{sc} \to H/Z \to 1.$$

This gives rise to the differential $\delta' : H^1(k, H/Z) \to H^2(k, Z^{sc})$. It is clear that δ is the composition of δ' and the natural map $H^2(k, Z^{sc}) \to H^2(k, Z)$. We conclude that $\alpha - \alpha'$ belongs to the image of this last map, and hence goes to 0 in $H^2_{ab}(k, L)$. QED

For a connected reductive k-lien L one defines a map

$$\mathrm{ab} : H^2(k, L) \to H^2_{ab}(k, L)$$

as follows. By property (1.4) the lien L is representable. Choose $\alpha \in H^2(k, L)'$, then define $\mathrm{ab}(x) \in H^2_{ab}(k, L)$ as the image of $x - \alpha \in H^2(k, Z)$. By the previous lemma this does not depend on the choice of the neutral element α, hence the map ab is well defined. This definition is in fact designed to be used in the case when k is a local or a number field.

Proposition 9.3.2 *Let L be a connected reductive k-lien, where k is a local field of characteristic 0 or a number field. Then*

$$H^2(k, L)' = \mathrm{ab}^{-1}(0).$$

Proof. In view of Lemma 9.3.1 we have to show that if $\alpha \in H^2(k, L)'$, $x \in H^2(k, L)$, and $x - \alpha \in H^2(k, Z)$ goes to 0 in $H^2_{ab}(k, L)$, then $x \in H^2(k, L)'$. By (9.5) $x - \alpha$ is in the image of $H^2(k, Z^{sc})$. It is well known that the map δ' is surjective in our case ([PR], 6.5, Thm. 20). Then $x - \alpha \in Im(\delta)$, hence by property (1.3) x is neutral. QED

9.4 Hasse principle for non-abelian H^2

We keep the notation of the preceding section. Let us denote H^{tor} by T. Let t be the natural map $H^2(k, L) \to H^2(k, T)$ (cf. [Bor 93], 1.7).

Theorem 9.4.1 (Borovoi) *Let L be a connected reductive k-lien over a number field k. An element $\eta \in H^2(k, L)$ is neutral if and only if its restriction to $H^2(k_v, L)$ is neutral for all the archimedean places v of k, and its image $t(\eta) \in H^2(k, T)$ is 0.*

Proof. By the previous proposition it is enough to work with the abelianized version $H^2_{ab}(k, L)$. Let H be a k-form of the underlying $\bar{k}$-group of L.

The exact sequence

$$1 \to H^{ss} \to H \to T \to 1$$

gives rise to the exact sequence of abelian k-groups

$$1 \to Z^{ss} \to Z \to T \to 1.$$

This in turn gives rise to the exact sequence of two-term complexes of abelian k-groups

$$1 \to (Z^{sc} \to Z^{ss}) \to (Z^{sc} \to Z) \to (1 \to T) \to 1.$$

Note that the complex $Z^{sc} \to Z^{ss}$ is quasi-isomorphic to the one-element complex consisting of μ in degree 0, where the finite abelian k-group μ is the fundamental group of H^{ss}. Applying the hypercohomology functor $\mathbb{H}^*(k, \cdot)$ we get the exact sequence of abelian groups

$$H^1(k, T) \to H^3(k, \mu) \to H^2_{ab}(k, L) \to H^2(k, T).$$

One checks that the composition of the third arrow with ab is the natural map t. The theorem now follows from this sequence and two well known facts: the map

$$H^3(k, \mu) \longrightarrow \prod_{v \in \Omega_\infty} H^3(k_v, \mu)$$

is an isomorphism [Milne, ADT]; and the map

$$H^1(k, T) \longrightarrow \prod_{v \in \Omega_\infty} H^1(k_v, T)$$

is surjective ([PR], 7.3, Cor. 2). QED

9.5 Descent on homogeneous spaces

The main result of this chapter is the following theorem.

Theorem 9.5.1 (Borovoi) *The Manin obstruction associated with* ℬ(X) *is the only obstruction to the Hasse principle for homogeneous spaces X of simply connected semisimple groups with connected stabilizers.*

More generally, the theorem is true for any connected linear group. When the stabilizer $\overline{H}$ is not connected but the natural extension of the group of connected components $\overline{H}/\overline{H}_0$ by $\overline{H}_0^{tor}$ is abelian, and $G^{ss} = G^{sc}$, the theorem is still true [Bor 96]. It is not clear whether the theorem holds for finite non-abelian stabilizers.

Proof of theorem. Our main goal is to prove that the Springer class η_X is neutral. To do so we want to show that $t(\eta_X) \in H^2(k, T)$ is the obstruction to the existence of universal X-torsors under the torus $T = H^{tor}$.

Using the cohomological triviality of commutative unipotent groups one shows that a 2-cocycle is neutral if and only if it has neutral image after passing to the quotient by the unipotent radical (see [Bor 93], Prop. 4.1). Therefore we can assume without loss of generality that $L = L_X$ is a connected reductive k-lien. Let H be a connected reductive k-group such that $L = lien(H)$ (property (1.4)).

Let G be the semisimple simply connected group acting transitively on X (on the right). We shall use the following known facts about such groups: $\overline{k}[G]^*/\overline{k}^* = 1$ (Rosenlicht's lemma), $Pic(\overline{G}) = 0$ (see [Sansuc 81], 6.5, 6.9).

Fix a point $\overline{x}_0 \in X(\overline{k})$. Let $f_0 : \overline{G} \to \overline{X}$ be the morphism given by $g \mapsto \overline{x}_0 g$ (this morphism is only defined over $\overline{k}$). Let $\overline{H} \subset \overline{G}$ be the stabilizer of $\overline{x}_0$. This is a closed algebraic subgroup of G defined over $\overline{k}$. The map $f_0 : \overline{G} \to \overline{X}$ makes $\overline{G}$ a left $\overline{X}$-torsor under $\overline{H}$. Then we can apply the exact sequence (2.7):

$$1 \to \overline{k}[X]^*/\overline{k}^* \to \overline{k}[G]^*/\overline{k}^* \to \hat{T} \to Pic(\overline{X}) \to Pic(\overline{G}).$$

This gives a natural isomorphism of abelian groups $\hat{T} \to Pic(\overline{X})$; let us call it λ. The same exact sequence also shows that $\overline{k}[X]^* = \overline{k}^*$. This implies that the adjunction morphism of $Spec(k)$-sheaves $T \to p_* p^* T$ is an isomorphism ($p : X \to Spec(k)$ being the structure morphism).

Let $\overline{Z} \to \overline{X}$ be the $\overline{X}$-torsor under $\overline{T}$, which is the push-forward of $f_0 : \overline{G} \to \overline{X}$ with respect to the map $\overline{H} \to \overline{T}$. It follows by functoriality of the displayed sequence above that $\texttt{type}(\overline{Z}) = \lambda$.

Let us prove that λ is an isomorphism of Γ_k-modules. It is enough to show that the type of $^\gamma\overline{Z} \to \overline{X}$ is λ, for all $\gamma \in \Gamma_k$. The action of Γ_k amounts to replacing $\overline{x}_0$ by $^\gamma\overline{x}_0$. More generally, let $\overline{x}_1$ be another $\overline{k}$-point of X. Choose $\overline{g}_0 \in G(\overline{k})$ such that $\overline{x}_0 \overline{g}_0 = \overline{x}_1$; then the stabilizer of $\overline{x}_1$ is $\overline{H}' = \overline{g}_0^{-1} \overline{H} \overline{g}_0$. This defines a $\overline{k}$-isomorphism between $\overline{H}$ and $\overline{H}'$, well defined up to a conjugation in $\overline{H}$. Thus the toric parts of $\overline{H}$ and of $\overline{H}'$ are canonically isomorphic, and so $\overline{H}'^{tor} = \overline{T}$. Let $f_1 : \overline{G} \to \overline{X}$ be the morphism given by $g \mapsto \overline{x}_1 g$. Let $\overline{Z}' \to \overline{X}$ be the $\overline{X}$-torsor under $\overline{T}$ obtained as the push-forward of the torsor $f_1 : \overline{G} \to \overline{X}$ under $\overline{H}'$. Moreover, the left translation of $\overline{G}$ by $\overline{g}_0$, $g \mapsto \overline{g}_0 g$, gives rise to an isomorphism of $\overline{X}$-torsors f_0 and f_1 (with respect to the isomorphism $\overline{H} \to \overline{H}'$ given by conjugation by $\overline{g}_0$). Hence $\overline{Z}$ and $\overline{Z}'$ are also isomorphic as $\overline{X}$-torsors under $\overline{T}$. This proves that λ does not depend on the choices made, and is an isomorphism of Γ_k-modules.

Thus $e(X) := \partial(\lambda)$, where ∂ is the differential coming from the spectral sequence of Ext's, is the obstruction to the existence of universal X-torsors of type λ (see Section 2.3). According to [Giraud], V.3.2.1 the second (abelian) cohomology class $e(X)$ is represented by the gerb $\mathcal{G}_\lambda$ of (universal) torsors of type λ, whose sections $\mathcal{G}_\lambda(K)$ over an extension $k \subset K$ are the isomorphism classes of X_K-torsors under T_K of type λ.

On the other hand, suppose that we are given a G_K-morphism of a right K-torsor Y_X under G_K to X_K. We can associate to it the X_K-torsor under T_K in the following natural way. Let $H_K = Aut_{G_K}(Y_K/X_K)$; this is an algebraic K-subgroup of G_K such that Y_K is an X_K-torsor under H_K (see Section 9.2). We associate to Y_K its push-forward with respect to the map $H_K \to T_K$. Our previous computation shows that the type of this torsor is λ (we check this over $\overline{k}$). Thus we get a morphism of gerbs $\mathcal{G}_X \to \mathcal{G}_\lambda$, where $\mathcal{G}_X$ is the gerb of liftings X_K to a K-torsor under G_K (the gerb of the homogeneous space X). The gerb $\mathcal{G}_\lambda$ is bound by the k-lien $p_*p^*T = T$ (we refer to [DM] or [Giraud] for the definition of the lien of a gerb). The gerb $\mathcal{G}_X$ is bound by H. The morphism of liens induced by $\mathcal{G}_X \to \mathcal{G}_\lambda$ is the natural surjection $H \to T$. This is enough to conclude that $t(\eta_X) = e(X)$.

By virtue of Proposition 6.1.4 we now obtain from the assumption of the theorem that $t(\eta_X) = 0$. By assumption $X(k_v) \neq \emptyset$ for all v, hence η_X is neutral everywhere locally by property (2.1). Theorem 9.4.1 then implies that η_X is neutral.

Therefore there exist a right k-torsor Y under G and a morphism $Y \to X$ compatible with the action of G. This also defines a k-group H (the group of G-automorphisms of $Y \to X$). The isomorphism class of Y is defined up to a twist by a k-torsor under H. Choose some points $P_v \in X(k_v)$ for all archimedean places v. Consider the classes of the fibres $[Y_{P_v}] \in H^1(k_v, H)$. It is known that the map

$$H^1(k, H) \to \prod_{v \in \Omega_\infty} H^1(k_v, H)$$

is surjective for any connected k-group H ([PR], 6.5, Prop. 17). Take a k-torsor under H whose class restricts to $[Y_{P_v}]$ for all archimedean places v. Twisting by this torsor we ensure that $Y(k_v) \neq \emptyset$ for all archimedean places v. The statement of the theorem now follows from the well known fact that the Hasse principle with respect to the real places holds for k-torsors under semisimple simply connected groups (Kneser, Harder and Chernousov). QED

Comments

In Sections 9.1 and 9.2 we follow [HS], [FSS] and [Springer 66]. Sections 9.3 and 9.4 are entirely due to Borovoi [Bor 93]. The proof of a particular case of Borovoi's theorem in Section 9.5 seems to be new. Borovoi's original proof in [Bor 96] though more involved has the merit that it does not use gerbs, and hence does not depend on [Giraud]. Our proof emphasizes the ubiquity of universal torsors.

I cannot think of any reason why Borovoi's theorem that the Manin obstruction to the Hasse principle is the only one for homogeneous spaces with connected or abelian stabilizers should remain true in the case of finite non-abelian stabilizers. However, to build an actual counter-example seems to be a difficult task.

Borovoi also proves in [Bor 96] that the Brauer–Manin obstruction to weak approximation is the only one for homogeneous spaces of connected affine algebraic groups with connected or abelian stabilizers (in the latter case one has to assume that $G^{ss} = G^{sc}$).

An interesting computation of the Manin obstruction to the Hasse principle on homogeneous spaces can be found in [Bor 99]. The main result of that paper is a generalization of Theorems 6.2.1 and 6.2.2 based on a version of Poitou–Tate duality for the hypercohomology of two-term complexes. In the notation of the last section the complex used by Borovoi is the natural complex of Γ_k-modules $\hat{G} \to \hat{T}$.

References

[AKMMMP] S.Y. An, S.Y. Kim, D.C. Marshall, S.H. Marshall, W.G. Mc-Callum, and A.R. Perlis. Jacobians of genus one curves. Preprint.

[BF] C.L. Basile and T. A. Fisher. Diagonal cubic equations in four variables with prime coefficients, to appear.

[Beauville] A. Beauville. *Surfaces algébriques complexes. Astérisque* **54** (1978).

[BCSS] A. Beauville, J.-L. Colliot-Thélène, J.-J. Sansuc et Sir Peter Swinnerton-Dyer. Variétés stablement rationnelles non rationnelles. *Ann. Math.* **121** (1985) 283–318.

[BBD] A. Beilinson, J. Bernstein et P. Deligne. *Faisceaux pervers. Astérisque* **100** (1982).

[BSw] A.O. Bender and Sir Peter Swinnerton-Dyer. Solubility of certain pencils of curves of genus 1, and of the intersection of two quadrics in $\mathbf{P}^4$. *Proc. London Math. Soc.*, to appear.

[Birch] B.J. Birch. Forms in many variables. *Proc. Royal Soc. London* **265** Ser. A (1961–62) 245–263.

[BSD] B.J. Birch and H.P.F. Swinnerton-Dyer. Notes on elliptic curves. I. *J. reine angew. Math.* **212** (1963) 7–25.

[BS] A. Borel et J.-P. Serre. Théorèmes de finitude en cohomologie galoisienne. *Comm. Math. Helv.* **39** (1964) 111–164.

[Bor 93] M.V. Borovoi. Abelianization of the second nonabelian Galois cohomology. *Duke Math. J.* **72** (1993) 217–239.

[Bor 96] M.V. Borovoi. The Brauer–Manin obstructions for homogeneous spaces with connected or abelian stabilizer. *J. reine angew. Math.* **473** (1996) 181–194.

[Bor 98] M.V. Borovoi. *Abelian Galois cohomology of reductive groups.* Memoirs AMS **626** (1998).

[Bor 99] M.V. Borovoi. A cohomological obstruction to the Hasse principle for homogeneous spaces. *Math. Annalen* **314** (1999) 491–504.

[BLR] S. Bosch, W. Lütkebohmert and M. Raynaud. *Néron models.* Springer-Verlag, 1990.

[BLM] A. Bremner, D.J. Lewis and P. Morton. Some varieties with points only in a field extension. *Arch. Math.* **43** (1984) 344–350.

[Cassels 62] J.W.S. Cassels. Arithmetic on curves of genus 1. III. The Tate–Shafarevich and Selmer Groups. *Proc. London Math. Soc.* **12** (1962) 259–296; IV. Proof of the Hauptvermutung. *J. reine angew. Math.* **211** (1962) 95–112.

[Cassels 85a] J.W.S. Cassels. The arithmetic of certain quartic curves. *Proc. Royal Soc. Edinburgh* **100A** (1985) 201–218.

[Cassels 85b] J.W.S. Cassels. Precise results in the arithmetic of curves of higher genera. *Uspekhi Mat. Nauk* **40** (1985) 43–47. (Russian)

[Cassels 98] J.W.S. Cassels. Second descents for elliptic curves. *J. reine angew. Math.* **494** (1998) 101–127.

[CFl] J.W.S. Cassels and E.V. Flynn. *Prolegomena to a middlebrow arithmetic of curves of genus 2.* LMS Lecture Notes Series **230**, Cambridge Univ. Press, 1996.

[CF] J.W.S. Cassels and A. Fröhlich eds. *Algebraic number theory.* Proc. Conf. Brighton. Thompson, 1967.

[Châtelet] F. Châtelet. Points rationnels sur certaines courbes et surfaces cubiques. *Enseign. Math.* **5** (1959) 153–170.

[C86] J.-L. Colliot-Thélène. Surfaces cubiques diagonales. In: *Séminaire de Théorie des Nombres de Paris 1984–1985.* Progress in Math. **63**, Birkhäuser, 1986, 51–66.

[C87] J.-L. Colliot-Thélène. Arithmétique des variétés rationnelles et problèmes birationnels. In: *Proc. Int. Cong. Math. Berkeley, 1986*, vol. 1 (1987) 641–653.

[C90] J.-L. Colliot-Thélène. Surfaces rationnelles fibrées en coniques de degré 4. In: *Séminaire de Théorie des Nombres de Paris 1988–1989.* Progress in Math. **91**, Birkhäuser, 1990, 43–55.

[C92] J.-L. Colliot-Thélène. L'arithmétique des variétés rationnelles. *Ann. Fac. Sci. Toulouse* **1** (1992) 51–73.

[C95] J.-L. Colliot-Thélène. L'arithmétique du groupe de Chow des zéro-cycles. *J. théorie des nombres de Bordeaux* **7** (1995) 51–73.

[C98] J.-L. Colliot-Thélène. The Hasse principle in a pencil of algebraic varieties. *Contemporary Math.* **210** (1998) 19–39.

[C99] J.-L. Colliot-Thélène. Conjectures de type local–global sur l'image des groupes de Chow dans la cohomologie étale. *Proc. Symp. Pure Math.* **67** (1999) 1–12.

[CCS] J.-L. Colliot-Thélène, D. Coray et J.-J. Sansuc. Descente et principe de Hasse pour certaines variétés rationnelles. *J. reine angew. Math.* **320** (1980) 150–191.

[CKS] J.-L. Colliot-Thélène, D. Kanevsky et J.-J. Sansuc. Arithmétique des surfaces cubiques diagonales. In: Lecture Notes Math. **1290**, Springer-Verlag, 1985, 1–108.

[CP] J.-L. Colliot-Thélène and B. Poonen. Algebraic families of nonzero elements of Tate–Shafarevich groups. *J. Amer. Math. Soc.* **13** (1999) 83–99.

[CS89] J.-L. Colliot-Thélène and P. Salberger. Arithmetic on some singular cubic hypersurfaces. *Proc. London Math. Soc.* **58** (1989) 519–549.

[CS77] J.-L. Colliot-Thélène et J.-J. Sansuc. La R-équivalence sur les tores. *Ann. Sci. École Norm. Sup.* **10** (1977) 175–230.

[CS80] J.-L. Colliot-Thélène et J.-J. Sansuc. La descente sur les variétés rationnelles. In: *Journées de géométrie algébrique d'Anger*, A. Beauville, ed. Sijthof and Noordhof, 1980, 223–237.

[CS87a] J.-L. Colliot-Thélène et J.-J. Sansuc. La descente sur les variétés rationnelles, II. *Duke Math. J.* **54** (1987) 375–492.

[CS87b] J.-L. Colliot-Thélène and J.-J. Sansuc. Principal homogeneous spaces under flasque tori: applications. *J. Algebra* **106** (1987) 148–205.

[CSS87] J.-L. Colliot-Thélène, J.-J. Sansuc and Sir Peter Swinnerton-Dyer. Intersections of two quadrics and Châtelet surfaces. *J. reine angew. Math.* **373** (1987) 37–107, **374** (1987) 72–168.

[CS92] J.-L. Colliot-Thélène et A.N. Skorobogatov. Approximation faible pour les intersections de deux quadriques en dimension 3. *C. R. Acad. Sci. Paris* **314** (1992) 127–132.

[CS00] J.-L. Colliot-Thélène and A.N. Skorobogatov. Descent on fibrations over $\mathbf{P}_k^1$ revisited. *Math. Proc. Camb. Phil. Soc.* **128** (2000) 383–393.

[CSS97] J.-L. Colliot-Thélène, A.N. Skorobogatov and Sir Peter Swinnerton-Dyer. Double fibres and double covers: paucity of rational points. *Acta Arithm.* **79** (1997) 113–135.

[CSS98a] J.-L. Colliot-Thélène, A.N. Skorobogatov and Sir Peter Swinnerton-Dyer. Rational points and zero-cycles on fibred varieties: Schinzel's hypothesis and Salberger's device. *J. reine angew. Math.* **495** (1998), 1–28.

[CSS98b] J.-L. Colliot-Thélène, A.N. Skorobogatov and Sir Peter Swinnerton-Dyer. Hasse principle for pencils of curves of genus one whose Jacobians have rational 2-division points. *Inv. Math.* **134** (1998), 579–650.

[CM] D.F. Coray and C. Manoil. On large Picard groups and the Hasse principle for curves and K3 surfaces. *Acta Arith.* **76** (1996), 165–189.

[CLSS] D.F. Coray, D.J. Lewis, N.I. Shepherd-Barron and Sir Peter Swinnerton-Dyer. Cubic threefolds with six double points. In: *Number theory in progress*. K. Győry, H. Iwaniec, J. Urbanowicz, eds. Walter de Gruyter, 1999, 63–74.

[Cremona] J.E. Cremona. *Algorithms for modular elliptic curves*. Cambridge Univ. Press, 1992.

[DM] P. Deligne and J. Milne. Appendix to: *Tannakian categories*. In: Lecture Notes Math. **900**, 1982, Springer-Verlag, 220–226.

[D] J.-C. Douai. Cohomologie galoisienne des groupes semisimples définis sur les corps globaux. *C. R. Acad. Sci. Paris* **275** (1975) 1077–1080.

[E] T. Ekedahl. An effective version of Hilbert's irreducibility theorem. In: *Séminaire de Théorie des Nombres de Paris 1988–1989*. Progress in Math. **91**, Birkhäuser, 1990, 242–249.

[FSS] Y.Z. Flicker, C. Scheiderer and R. Sujatha. Grothendieck's theorem on non-abelian H^2 and local–global principles. *J. Amer. Math. Soc.* **11** (1998) 731–750.

[GPZ] J. Gebel, A. Pethő and H. G. Zimmer. On Mordell's equation. *Compositio Math.* **110** (1998) 335–367.

[GM] S.I. Gelfand and Yu.I. Manin. *Methods of homological algebra*. Nauka, 1988 (Russian). English translation: *Homological algebra*. Encyclopaedia Math. Sci. **38**, Springer-Verlag, 1994.

[Giraud] J. Giraud. *Cohomologie non abélienne*. Springer-Verlag, 1971.

[Grothendieck 57] A. Grothendieck. Sur quelques points d'algèbre homologique. *Tôhoku Math. J.* **9** (1957) 119–221.

[Grothendieck, SGA 1] A. Grothendieck. *Revêtements étales et groupe fondamental*. Séminaire de Géométrie Algébrique du Bois Marie 1960/61. Lecture Notes Math. **224**, Springer-Verlag, 1971.

[Grothendieck, SGA 4] A. Grothendieck. *Théorie des topos et cohomologie étale des schémas*. Séminaire de Géométrie Algébrique du Bois Marie 1963/64. Lecture Notes Math. **269, 270, 305**, Springer-Verlag, 1972–1973.

[Grothendieck 68] A. Grothendieck. Le groupe de Brauer. I, II, III. In: *Dix exposés sur la cohomologie des schémas*. A. Grothendieck, N. H. Kuipers, eds. North-Holland, 1968, 46–188.

[Grothendieck 84] A. Grothendieck. Esquisse d'un programme. (1984).

In: *Geometric Galois actions, I*. L. Schneps, P. Lochak, eds. London Math. Soc. Lecture Notes **242**, Cambridge Univ. Press, 1997, 49–58.

[GMY] V. Guletskiĭ, G.L. Margolin and V. Yanchevskiĭ. Presentation of two-torsion part of the Brauer groups of curves by quaternion algebras. *Doklady NAN Belorus* **41** (1997) 4–8. (Russian)

[H94] D. Harari. Méthode des fibrations et obstruction de Manin. *Duke Math. J.* **75** (1994) 221–260.

[H95] D. Harari. Principe de Hasse et approximation faible sur certaines hypersurfaces. *Ann. Fac. Sci. Toulouse* **4** (1995) 731–762.

[H96] D. Harari. Obstructions de Manin "transcendantes". In: *Séminaire de Théorie des Nombres de Paris 1993–1994*. S. David ed. Cambridge Univ. Press, 1996, 75–87.

[H97] D. Harari. Flèches de spécialisation en cohomologie étale et applications arithmétiques. *Bull. Soc. Math. France* **125** (1997) 143–166.

[H] D. Harari. Weak approximation and non-abelian fundamental groups. *Ann. Sci. École Norm. Sup.* **33** (2000) 467–484.

[HS] D. Harari and A.N. Skorobogatov. Non-abelian cohomology and rational points. *Compositio Math.*, to appear.

[Hartshorne] R. Hartshorne. *Algebraic geometry*. Springer-Verlag, 1977.

[HB] D.R. Heath-Brown. The solubility of diagonal cubic diophantine equations. *Proc. London Math. Soc.* (3) **79** (1999) 241–259.

[Iversen] B. Iversen. Brauer group of a linear algebraic group. *J. Algebra* **42** (1976) 295–301.

[Iskovskih] V.A. Iskovskih. A counter-example to the Hasse principle for a system of two quadratic forms in five variables. *Mat. Zametki* **10** (1971) 253–257 (Russian). English translation: *Math. Notes* **10** (1971) 575–577.

[KL] N.M. Katz and S. Lang. Finiteness theorems in geometric classfield theory. *Enseign. Math.* **27** (1981) 285–314.

[KS] B.È. Kunyavskiĭ and A.N. Skorobogatov. Weak approximation in algebraic groups and homogeneous spaces. *Contemporary Math.* **131** (1992) 447–451.

[Lang 54] S. Lang. Some applications of the local uniformization theorem. *Amer. J. Math.* **76** (1954) 362–374.

[Lang 56] S. Lang. Algebraic groups over finite fields. *Amer. J. Math.* **78** (1956) 555–563.

[Lang, AV] S. Lang. *Abelian varieties*. Interscience, 1959; 2nd ed. Springer-Verlag, 1983.

[Lang, ANT] S. Lang. *Algebraic number theory*. Springer-Verlag, 1986.

[LW] S. Lang and A. Weil. Number of points of varieties over finite fields. *Amer. J. Math.* **76** (1954) 819–827.

[Manin 70] Yu.I. Manin. Le groupe de Brauer–Grothendieck en géométrie diophantienne. In: *Actes Congrès Int. Math. Nice, 1970*, tome **I**, Gauthier-Villars, 1971, 401–411.

[Manin, L] Yu.I. Manin. *A course in mathematical logic.* Springer-Verlag, 1977.

[Manin, CF] Yu.I. Manin. *Cubic forms.* Nauka, 1972 (Russian). English translation: North-Holland, 2nd ed., 1986.

[MT] Yu.I. Manin and M.A. Tsfasman. Rational varieties: algebra, geometry, arithmetic. *Uspekhi Mat. Nauk* **41** (1986) 43–94. (Russian) English translation: *Russian Math. Surveys* **41** (1986) 51–116.

[M92] B. Mazur. The topology of rational points. *J. Experimental Math.* **1** (1992) 35–45.

[M94] B. Mazur. Questions of decidability and undecidability in number theory. *J. Sym. Logic* **59** (1994) 353–371.

[M95] B. Mazur. Speculations about the topology of rational points: an up-date. *Astérisque* **228** (1995) 165–181.

[MSS] J.R. Merriman, S. Siksek and N.P. Smart. Explicit 4-descents on an elliptic curve. *Acta Arith.* **77** (1996) 385–404.

[Milne, ADT] J.S. Milne. *Arithmetic duality theorems.* Persp. Math. **1**, Academic Press, 1986.

[Milne, EC] J.S. Milne. *Étale cohomology.* Princeton Univ. Press, 1980.

[Mumford, AV] D. Mumford. *Abelian varieties.* Tata Inst. Fund. Res. Studies in Math. **5**, 1970.

[Mumford, T] D. Mumford. *Tata lecture on Theta II.* Progress in Math. **43**, Birkhäuser, 1984.

[Mumford, GIT] D. Mumford, J. Fogarty and F. Kirwan. *Geometric invariant theory.* 3rd ed., Springer-Verlag, 1994.

[N] H. Nishimura. Some remark on rational points. *Mem. Coll. Sci. Kyoto*, Ser. A **29** (1955) 189–192.

[PR] V.P. Platonov and A.S. Rapinchuk. *Algebraic groups and number theory.* Nauka, 1991 (Russian). English translation: Academic Press, 1994.

[R] K. Rubin. Tate–Shafarevich groups and L-functions of elliptic curves with complex multiplication. *Inv. Math.* **89** (1987) 527–559.

[Saito] S. Saito. Some observations on motivic cohomology of arithmetic schemes. *Inv. Math.* **98** (1989) 371–404.

[Sl88] P. Salberger. Zero-cycles on rational surfaces over number fields. *Inv. Math.* **91** (1988) 505–524.

[Sl89] P. Salberger. On the arithmetic of certain intersections of two quadrics. *J. reine angew. Math.* **394** (1989) 159–167.

[SS] P. Salberger and A.N. Skorobogatov. Weak approximation for surfaces defined by two quadratic forms. *Duke Math. J.* **63** (1991) 517–536.

[Sansuc 81] J.-J. Sansuc. Groupe de Brauer et arithmétique des groupes algébriques linéaires sur un corps de nombres. *J. reine angew. Math.* **327** (1981) 12–80.

[Sansuc 82] J.-J. Sansuc. Descente et principe de Hasse pour certaines variétés rationnelles. In: *Séminaire de Théorie des Nombres de Paris 1980–1981.* Birkhäuser, 1982, 253–271.

[Sansuc 87] J.-J. Sansuc. Principe de Hasse, surfaces cubiques et intersections de deux quadriques. In: *Journées arithmétiques de Besançon, 1985. Astérisque* **147–148** (1987) 183–207.

[SW] P. Sarnak and L. Wang. Some hypersurfaces in $\mathbf{P}^4$ and the Hasse-principle. *C. R. Acad. Sci. Paris* **321** (1995) 319–322.

[Sch] V. Scharaschkin. The Brauer–Manin obstruction for curves. Preprint, 1998.

[Serre, CG] J.-P. Serre. *Cohomologie galoisienne.* 5ème éd., Lecture Notes. Math. **5**, Springer-Verlag, 1994.

[Serre, GA] J.-P. Serre. *Groupes algébriques et corps de classes.* Hermann, 1959.

[Serre, CL] J.-P. Serre. *Corps locaux.* Hermann, 1968.

[Shafarevich] I.R. Shafarevich *et al.* *Algebraic surfaces. Trudy Mat. Inst. Steklov* **75** (1965) (Russian). English translation: *Proc. Steklov Inst. Math.* **75** (1967).

[SB92] N.I. Shepherd-Barron. The rationality of quintic Del Pezzo surfaces – a short proof. *Bull. London Math. Soc.* **24** (1992) 249–250.

[SB98] N.I. Shepherd-Barron. Arithmetic on 9-nodal cubic 3-folds. Preprint, 1998.

[Sh97] A. Shlapentokh. Diophantine definability over some rings of algebraic numbers with infinite number of primes allowed in denominator. *Inv. Math.* **129** (1997) 489–507.

[S90a] A.N. Skorobogatov. Arithmetic on certain quadric bundles of relative dimension 2, I. *J. reine angew. Math.* **407** (1990) 57–74.

[S90b] A.N. Skorobogatov. On the fibration method for proving the Hasse principle and weak approximation. In: *Séminaire de Théorie des Nombres de Paris 1988–1989.* Progress in Math. **91**, Birkhäuser, 1990, 205–219.

[S93] A.N. Skorobogatov. On a theorem of Enriques–Swinnerton-Dyer. *Ann. Fac. Sci. Toulouse* **2** (1993) 429–440.

[S96] A.N. Skorobogatov. Descent on fibrations over the projective line. *Amer. J. Math.* **118** (1996) 905–923.

[S99] A.N. Skorobogatov. Beyond the Manin obstruction. *Inv. Math.* **135** (1999) 399–424.

[Springer 66] T.A. Springer. Nonabelian H^2 in Galois cohomology. In: *Algebraic groups and discontinuous subgroups.* A. Borel, G.D. Mostow, eds. Proc. Symp. Pure Math. **9**, Amer. Math. Soc., 1966, 164–182.

[Springer 81] T.A. Springer. *Linear algebraic groups.* Progress in Math. **9**, Birkhäuser, 1981.

[SD72] H.P.F. Swinnerton-Dyer. Rational points on del Pezzo surfaces of degree 5. In: *Algebraic geometry* (Proc. Fifth Nordic Summer School in Math., Oslo 1970) Wolters–Noordhoff, 1972, 297–290.

[SD95] Sir Peter Swinnerton-Dyer. Rational points on certain intersections of two quadrics. In: *Abelian varieties*, W. Barth, K. Hulek and H. Lange eds. Walter de Gruyter, 1995, 273–292.

[SD96] Sir Peter Swinnerton-Dyer. *Diophantine equations: the geometric approach.* Jber. d. Dt. Math.-Verein. **98** (1996) 146–164.

[SD99] Sir Peter Swinnerton-Dyer. Rational points on some pencils of conics with 6 singular fibres. *Ann. Fac. Sci. Toulouse* **8** (1999) 331–341.

[SD00a] Sir Peter Swinnerton-Dyer. Arithmetic of diagonal quartic surfaces, II. *Proc. London Math. Soc.* **80** (2000) 513–544.

[SD00b] Sir Peter Swinnerton-Dyer. The solubility of diagonal cubic surfaces. *Ann. Sci. École Norm. Sup.*, to appear.

[V, AT] V.E. Voskresenskiĭ. *Algebraic tori.* Nauka, 1977 (Russian).

[V, AG] V.E. Voskresenskiĭ. *Algebraic groups and their birational transformations.* Transl. Math. Monographs **179**, Amer. Math. Soc., 1998.

[Wang] L. Wang. Brauer–Manin obstruction to weak approximation on abelian varieties. *Israel J. Math.* **94** (1996) 189–200.

[Weibel] C.A. Weibel. *An introduction to homological algebra.* Cambridge Univ. Press, 1994.

[Weil 83] A. Weil. *Number theory. An approach through history.* Birkhäuser, 1983.

[Weil 55] A. Weil. On the algebraic groups and homogeneous spaces. *Amer. J. Math.* **77** (1955) 493–512.

[YM] V. Yanchevskiĭ and G.L. Margolin. The Brauer groups and the torsion of local elliptic curves. *St Petersburg Math. J.* **7** (1996) 473–505.

Index

An italic page number indicates the primary reference.